Goat Keeping 101

And Their Place in the Homestead

⚬

MORGAN DESPIEGELAERE

I S B N 978-1-964165-70-7

Dedication

To my husband Jason, our son Declan, and our daughter Serena, for always being there with me by my side while we embark on many crazy journeys together on our family farm/homestead

Contents

Author's Note

Welcome to the world of goats! After spending many years teaching others how to care for goats and various other species as a Registered Veterinary Technician (RVT), I'm excited to finally share my knowledge and passion for these incredible animals with you. This book is a labor of love, created to help anyone who wants to better understand and appreciate goats. I hope you enjoy this journey as much as I have, and thank you for joining me in exploring the fascinating world of goat keeping!

Chapter One:
Introduction to Goats

Welcome to the World of Goats

Goats are one of the oldest domesticated animals in human history. They've been around for thousands of years, playing a significant role in the survival and development of various civilizations. From providing milk, meat, and fiber to acting as natural weed eaters and land clearers, goats have earned their place in cultures across the globe. Whether it's a subsistence farmer in Africa, a cheese maker in France, or a homesteader in North America, goats continue to be invaluable in various ecosystems and lifestyles.

For the modern homesteader or small-scale farmer, goats offer an incredible balance of practicality and charm. This chapter introduces you to the world of goat keeping and sets the stage for the journey ahead, offering a comprehensive look at why goats are such a fantastic addition to any homestead.

A Rich History of Goat Domestication

The domestication of goats began over 10,000 years ago, likely in the rugged hills of present-day Iran and Turkey. Goats were prized not only for their milk and meat but also for their hardy nature and ability to thrive in diverse environments. They became a staple in the agricultural practices of ancient civilizations, from the Nile Valley to the Mediterranean, and were depicted in everything from ancient art to religious texts.

Their adaptability is a key reason why goats are found in almost every part of the world today, from the mountains of Nepal to the grasslands of Texas.

Why Goats Are Perfect for Small-Scale Farming and Homesteads

When considering animals for your homestead, you might be tempted to think about chickens for eggs, cows for milk, or pigs for meat. But goats bring something unique to the table: they can fulfill multiple roles simultaneously while being manageable in terms of size, care, and resource needs.

- Adaptability: Goats can thrive in a variety of environments—from desert terrains to lush pastures. They are relatively low-maintenance animals that are incredibly resilient and can adapt to both hot and cold climates. Whether you're in an arid region or a temperate climate, goats can acclimate with the right care.

- Manageable Size: Unlike cows or horses, which can be difficult to house and feed on small homesteads, goats are of a manageable size. They don't require enormous barns or extensive fencing. A modest homestead can comfortably accommodate a small herd, making them ideal for families just starting with livestock.

- Intelligence and Social Structure: Goats are intelligent, curious creatures that can be trained, which is an advantage for managing them. Their natural herd instinct makes them social animals, and this social behavior contributes to their ease of care. Goats thrive in groups, and with proper socialization and routine, they quickly adapt to the rhythms of farm life.

- Multiple Uses: One of the biggest reasons goats have remained popular for so long is their multifunctionality. A single goat can provide milk, meat, and fiber, and their manure is a fantastic fertilizer for your homestead. They're also known for their ability to control weeds and clear brush, making them natural landscapers. Their grazing habits help maintain pastures and contribute to better land management.

Goats and Sustainability

In today's world, sustainable farming and living are no longer luxuries—they are necessities. As climate change impacts traditional farming practices, there is a renewed interest in practices that can support local ecosystems while reducing reliance on commercial food systems. Goats play a vital role in sustainable agriculture for several reasons:

- Low Environmental Impact: Goats require far fewer resources compared to larger livestock like cattle. They eat less, take up less space, and their waste is more manageable, making them a sustainable option for small farms.

- Efficient Grazers: Goats have a unique ability to eat plants that other livestock ignore. This means they can clear invasive species, reduce fire hazards in overgrown areas, and improve the health of your pasture without the need for chemicals or heavy machinery. Their grazing behavior is beneficial for land management, as they consume shrubs and weeds that can otherwise crowd out more desirable plants.

- Manure for Fertilization: Goat manure is rich in nitrogen, potassium, and phosphorus, making it an excellent natural fertilizer for gardens and crops. By composting their waste, you can create a cycle of sustainability where your land continues to grow and improve over time.

- Milk, Meat, and Fiber: Raising goats can drastically reduce your reliance on grocery stores. Goat milk can be used for drinking, cheese making, and even soap production. Their meat is a lean, healthy alternative to beef or pork, and certain breeds like Angora or Cashmere goats provide valuable fibers for textiles. By raising goats, you're contributing to your family's self-sufficiency while reducing the environmental impact of commercial farming.

Economic Benefits of Goat-Keeping

From an economic perspective, goats are an investment that can start paying dividends almost immediately. The cost of raising goats is relatively low compared to larger livestock, yet they offer numerous financial benefits.

- Reducing Food Costs: For dairy enthusiasts, a couple of well-cared-for dairy goats can supply your household with enough milk for daily consumption, cheese, and yogurt production. This reduces your reliance on store-bought dairy and offsets the costs associated with commercial food.

- Income from Products: Many homesteaders find that goat products like milk, cheese, soap, and fiber can be sold at farmers' markets or to local communities. There's a growing demand for natural, farm-to-table products, and goats allow you to tap into this market.

- Brush Clearing and Agritourism: Some enterprising goat keepers rent out their herds for brush clearing services. Goats can help manage land by eating invasive plants, and the idea of "goat rental" has gained popularity, especially in fire-prone areas where reducing vegetation is critical. Additionally, agritourism, like "goat yoga" and educational farm tours, has become a novel way for farmers to engage with the public and generate income.

Personal Insights: My Experience as an RVT and Goat Keeper

After spending 11 years as a Registered Veterinary Technician (RVT) and raising many animals on my own farm, I can attest to the uniqueness and joy of working with goats. I've helped people care for everything from cats and dogs to horses and other livestock animals, but goats hold a special place in my heart. Their intelligence, curiosity, and individuality make them endlessly entertaining, and their care is both rewarding and relatively straightforward.

One of the most valuable lessons I've learned in my years of working with goats is that they require a balance of structure and flexibility. On one hand, routine is crucial. Goats, like many creatures, prefer a consistent routine. They thrive when they know when to expect feeding, milking, or even playtime. However, flexibility is equally important because goats, being as curious and mischievous as they are, often throw curveballs. They might escape their pen, eat something they shouldn't, or refuse to cooperate just because they can. It's all part of the adventure!

From a veterinary technician's standpoint, goats are relatively hardy creatures, but they still require proper care and attention to ensure their well-being. I've seen goats with preventable health issues simply because their owners weren't familiar with their needs. My goal is to make sure that anyone reading this book is equipped with the essential knowledge they need to avoid those problems. Simple things like ensuring goats have access to clean water, providing a balanced diet, and regularly checking their hooves can make a world of difference in their health and longevity.

The Enriching Experience of Goat Keeping

Caring for goats goes beyond the practical benefits of milk, meat, and fiber. It's an experience that can teach you valuable life lessons about care, responsibility, and commitment. Homesteaders are the stewards of the animals they raise, and with that comes a deep sense of connection and responsibility.

One of the most fulfilling aspects of goat keeping is the bond that develops between the animals and their caregivers. Goats, much like dogs or horses, have personalities. They remember you, greet you when you come into the pen, and even develop individual preferences for certain foods, routines, or interactions. Some goats are more independent, while others are affectionate and enjoy being scratched behind the ears or sitting with you in the barn.

Goat keeping is also a fantastic family endeavor. Children can learn responsibility by helping with daily chores like feeding or milking, and they develop an appreciation for animals and the environment around them. On a homestead, there's always something to do—whether it's mucking out a pen, feeding a hungry kid, or simply watching the herd graze contentedly in the pasture. These tasks foster a sense of purpose and satisfaction that is hard to find elsewhere.

Raising goats encourages self-sufficiency, patience, and a deeper connection to nature. As you begin this journey, you'll find that the lessons learned in the barn or pasture will extend into other parts of your life. Goats will become a part of your family, and the experience of raising them brings joy, fulfillment, and an understanding of what it means to truly care for something beyond yourself.

A Final Word

Remember that goat keeping is not just about the animals themselves but also about the lifestyle, which is about building a homestead that values sustainability, responsibility, and harmony with the land. Whether you are raising goats for milk, meat, or fiber or simply for the joy of their company, the journey will be as enriching as it is rewarding.

The following chapters will guide you through the essential aspects of goat keeping, from choosing the right breed to ensuring their health and well-being. Each section will build on the knowledge you've gained, ensuring that by the end of this book, you'll feel confident in your ability to care for and manage your herd. Welcome to the world of goats—let's get started!

Chapter Two:
Choosing the Right Goat Breed and Understanding Goat Behavior

Selecting the right goat breed for your homestead is one of the most important decisions you'll make as a goat keeper. Each breed has its own strengths, temperament, and specific needs, so finding the right fit for your environment and goals is key to building a healthy, productive herd. Beyond choosing a breed, understanding goat behavior—how they communicate, socialize, and interact with their environment—is essential for ensuring that you can manage and care for your goats effectively.

In this chapter, we will explore the most popular goat breeds, focusing on their physical traits, temperaments, and purposes so you can make an informed decision. We will also dive into goat behavior, examining their social structures, communication methods, and tips for managing their behavior in ways that enhance the homesteading experience.

Popular Goat Breeds by Purpose

When it comes to goat breeds, the diversity is astounding. Goats can be bred for milk, meat, fiber, or even as pets. Your choice of breed should reflect your goals—whether you're looking for high milk production, sustainable fiber, or a hardy meat source. Here's an overview of the most common breeds categorized by purpose.

Dairy Breeds

- <u>Nubian</u>: Known for their long, floppy ears and gentle temperament, Nubians are a popular choice for dairy. They produce high-butterfat milk, which makes it perfect for cheese making. Nubians are also more vocal than other breeds, which can be charming or annoying, depending on your preferences. They adapt well to various climates, making them a versatile breed.

- <u>Saanen</u>: Saanens are often referred to as the "Holsteins of the goat world" due to their prolific milk production. This breed produces large quantities of milk with a relatively low butterfat content. Saanens are large, white goats with a calm temperament, making them great for first-time goat owners. However, they can be sensitive to heat, so they thrive best in cooler climates.

- <u>Alpine</u>: These medium-to-large goats are incredibly hardy and adaptable to different environments. They are excellent milk producers, offering a good balance between quantity and butterfat. Alpines are alert and friendly, though they may be a bit more independent than other dairy breeds.

- LaMancha: Recognized by their unique "gopher" or "elf" ears, LaManchas are known for their high milk production and friendly, calm demeanor. They are highly adaptable to both hot and cold climates, making them an ideal choice for a range of homesteaders. Their milk is high in butterfat, and they are generally easy to handle and maintain.

Meat Breeds

- Boer: Originating from South Africa, Boer goats are the gold standard in meat production. They are large, muscular, and grow rapidly, making them ideal for commercial meat production. Boer goats are also docile and easy to manage, though they can be a bit lazy compared to other breeds. They adapt well to a variety of climates, though they may need more feed and care than some of the smaller breeds.

- Kiko: Kiko goats are a hardy meat breed developed in New Zealand to thrive in tough conditions with minimal input. They are highly resistant to parasites and require less maintenance compared to other breeds. Kikos are known for their fast growth rates and good maternal instincts. If you're looking for a low-maintenance, robust meat goat, Kikos are an excellent option.

- Spanish Goats: These goats are well-suited for brush control and meat production. They are hardy and can survive in difficult environments, making them ideal for homesteaders with rugged or scrubby land. Spanish goats have strong survival instincts, making them resistant to parasites and diseases.

Fiber Breeds

- <u>Angora</u>: If you're interested in producing fiber, Angora goats are your best bet. They are the source of mohair, a luxurious, silky fiber. Angoras require more care than meat or dairy breeds, as their coats need regular shearing and attention. They are generally docile and can be kept with other animals, but their fleece does make them more susceptible to parasites.

- <u>Cashmere</u>: Technically, many different breeds can produce cashmere, but goats specifically bred for this fine undercoat are known for producing high-quality fiber. Cashmere goats are hardy and do not require as much maintenance as Angoras, though their fiber does need careful harvesting. They're best suited for colder climates, where their thick coats grow in abundance.

Pet and Companion Breeds

- <u>Pygmy Goats</u>: Small, friendly, and great with kids (both human and goat!), Pygmy goats are ideal as pets. They don't produce much milk or meat, but their playful nature makes them a favorite among homesteaders looking for low-maintenance, entertaining companions. Pygmy goats adapt well to various environments and are relatively easy to care for, though they are prone to becoming overweight if overfed.

- <u>Nigerian Dwarf</u>: While Nigerian Dwarfs are primarily known for their small stature and friendly disposition, they are also great little milkers. They produce surprisingly large amounts of high-butterfat milk for their size. Nigerian Dwarfs are easy to manage, making them popular among both hobby farmers and serious goat breeders. They do well in both cold and warm climates, making them a versatile choice for different homesteads.

Choosing the Right Breed for Your Homestead

Now that you're familiar with the most common goat breeds, how do you choose the right one for your farm? Here are some key factors to consider:

- Climate: Some goats thrive in cooler climates (like Saanens), while others can handle heat better (such as Nubians or Kikos). Consider the temperature extremes in your region and select a breed that will be comfortable throughout the year.

- Purpose: Do you want milk, meat, fiber, or a companion? For milk, Nubians and Alpines are excellent choices. For meat, Boers and Kikos will serve you well. If fiber is your goal, Angoras are your best bet.

- Land: Goats are natural foragers, but not all land is created equal. If your land is rugged or brush-heavy, a hardy breed like Spanish or Kiko goats might be ideal. For more open pastures, dairy breeds like Saanens or Alpines can graze more efficiently.

- Management Time: Some goats, like Angoras, require more hands-on care due to their fleece. Others, like Kikos, are more self-sufficient and less prone to illness or parasite infestations. Consider how much time you have to dedicate to grooming, feeding, and overall care.

Understanding Goat Behavior:
Social Structures and Herd Dynamics

Goats are highly social creatures. They thrive in a herd environment, and their social interactions are central to their well-being. Understanding goat social structures can help you manage your herd more effectively and prevent behavioral problems.

- Hierarchy: Goats have a strict social hierarchy. Each herd will have a dominant goat—often referred to as the herd queen—who decides where and when the herd moves. This hierarchy isn't limited to does; bucks and wethers (castrated males) also establish their own pecking order.

- Dominant and Submissive Behavior: Dominant goats will push and shove others, often taking the best food and water spots. Submissive goats, on the other hand, will yield to these more aggressive individuals. It's important to recognize these behaviors when managing feeding times or introducing new goats to the herd. Keeping your herd size manageable and ensuring enough food and water for all goats will help reduce conflict.

How Goats Communicate:
Vocalizations, Body Language, and Physical Cues

Goats are highly communicative, using a combination of vocalizations, body language, and physical cues to express their needs, emotions, and intentions.

- Vocalizations: Goats bleat for many reasons—hunger, loneliness, excitement, or discomfort. Each goat's voice is unique, and you'll quickly learn to recognize the difference between a distressed bleat and a contented one. Nubians, in particular, are known for their loud, frequent vocalizations, while other breeds, like Alpines, are more reserved.

- Body Language: Goats often express themselves through posture. A goat standing tall with raised hair is asserting dominance, while one that's hunched or standing with lowered ears may be feeling submissive or unwell. Head-butting is a common form of communication among goats, often used to establish dominance or play.

- Physical Cues: Pay attention to tail wagging, head shaking, and even subtle ear movements. Goats will flick their ears toward sounds or stimuli they're interested in, and a wagging tail can indicate happiness or excitement, much like in dogs.

Managing Goat Behavior

While goats are intelligent and curious, their behavior can sometimes become challenging. Here are strategies for managing common behavioral issues:

- Socializing Goats: Goats are happiest when they have companionship, whether from other goats or even other animals like sheep or horses. Isolation can lead to stress, so always aim to keep goats in pairs or small herds.

- Reducing Stress: Goats can become stressed from sudden changes, such as new environments, predators, or handling. To reduce stress, maintain a consistent routine, introduce new goats gradually, and ensure that your herd feels secure with proper fencing and shelter.

- **Dealing with Aggression**: Aggression in goats is often a result of hierarchy disputes. Separate aggressive goats temporarily if fights become too intense, and provide multiple feeding and watering stations to avoid competition.

Training and Enrichment for Goats

Believe it or not, goats can be trained much like dogs. They respond well to consistent, positive reinforcement, and training can make managing your herd much easier.

- **Leash Training**: Goats can be trained to walk on a leash, which is useful for moving them around the homestead or during veterinary visits. Start with a soft halter and practice short walks, rewarding them with treats for cooperation.

- **Verbal Commands**: Goats can learn simple commands like "come," "stay," and "no." This training is particularly useful for milking goats, as it encourages them to be calm and cooperative during the process.

- **Enrichment**: Goats are intelligent and need mental stimulation to prevent boredom. Offer them climbing structures, toys, and different types of forage to keep them engaged. Enrichment reduces destructive behaviors like chewing on fences or climbing out of pens.

Conclusion

Choosing the right goat breed and understanding goat behavior is fundamental to successful goat keeping. By selecting a breed that fits your goals and climate and by learning to interpret and manage your goats' social and communicative behaviors, you'll set the foundation for a happy, healthy herd. Training, enrichment, and good management practices will not only ensure that your goats thrive but will also make your experience as a goat keeper more enjoyable and rewarding.

∽✲∽

Chapter Three:
Setting Up for Goats on the Homestead/Farm

Creating a safe and efficient environment for your goats is essential for their health, well-being, and your success as a goat keeper. Proper preparation ensures your goats are comfortable, healthy, and protected from potential hazards. In this chapter, we will walk you through the critical components of setting up a homestead or farm for goats, covering housing, fencing, pasture management, water systems, and more. Whether you are starting from scratch or upgrading your existing setup, thoughtful planning will make a world of difference in managing your herd smoothly.

Goat Housing: Building Safe, Durable Shelters

Your first priority when preparing for goats is to provide them with a secure shelter. Goats need protection from the elements—rain, snow, wind, and extreme heat—as well as predators. A well-built shelter also gives them a dry, comfortable place to sleep and rest.

Essentials of Goat Shelters

- Durability and Security: Goats are curious and clever; they can open latches, push doors, and even climb unstable structures. It's important to use strong materials—wood, metal, or plastic—and secure latches on doors to keep goats safe inside and predators out.

- Weather Protection: Depending on your region, your shelter must shield goats from heat, rain, or snow. A barn, shed, or three-sided shelter will work, provided it blocks prevailing winds and offers adequate cover.

- Space Requirements: Each goat should have about 15-20 square feet of indoor space, with additional outdoor space for exercise. Overcrowding can lead to stress and health issues, so it's crucial to provide enough room.

- Ventilation: Proper airflow prevents the buildup of moisture and ammonia from urine, both of which can cause respiratory problems. Windows or ventilation holes placed near the roofline will help keep the shelter dry and fresh.

- Bedding: Use straw, wood shavings, or hay as bedding material to keep the shelter warm and comfortable. Regularly clean out old bedding to prevent the buildup of parasites or mold.

Fencing: Securing Your Goats and Protecting Your Farm

When it comes to fencing, think of goats as escape artists. They are natural climbers, jumpers, and inquisitive explorers, which means your fencing needs to be secure, tall, and well-maintained.

Goat-Proof Fencing Guidelines

- <u>Height</u>: Fences should be at least 4-5 feet tall to prevent goats from jumping over. Larger breeds like Boers may need even higher fences.

- <u>Materials</u>: Woven wire or livestock panels are ideal because they resist climbing. Avoid barbed wire, as it can injure goats if they try to climb or push through it.

- <u>Posts and Bracing</u>: Set fence posts deeply into the ground (2-3 feet) to keep the fence sturdy. Corner posts should be well-braced to prevent sagging.

- <u>Electric Fencing</u>: Adding a single strand of electric wire along the top or bottom of your fence can be an effective deterrent against escape attempts. It also helps keep predators out.

- <u>Maintenance</u>: Regularly inspect fences for gaps, sagging wires, or damage. Goats will quickly find weak spots, so prompt repairs are essential.

Managing Pastures and Grazing

Goats are natural browsers rather than grazers, meaning they prefer shrubs, weeds, and leaves over grass. Proper pasture management ensures that your land stays healthy and your goats have access to diverse forage throughout the year.

Rotational Grazing

Rotational grazing is one of the best ways to manage pasture health. Divide your land into several smaller sections (paddocks) and rotate the goats between them. This allows vegetation to recover between grazing periods and helps prevent overgrazing.

Soil and Vegetation Health

- Prevent Overgrazing: Goats are notorious for eating vegetation down to the roots if left in one area too long. Rotating pastures preserves soil health and ensures that plants can regrow.

- Weed Management: Goats will eat many weeds and invasive plants that other livestock avoid. This helps keep pastures healthy and reduces the need for herbicides.

- Supplementary Feeding: In times of drought or during winter, pasture may not provide enough nutrition. Be prepared to supplement with hay, grains, and minerals when forage is scarce.

Water Management:
Keeping Your Goats Hydrated Year-Round

Access to clean water is essential for goat health. Goats need fresh water to regulate their body temperature, produce milk, and digest food properly. Dehydration can lead to serious health issues, so it's critical to ensure that water is always available.

Water Systems for Goats

- Buckets or Troughs: Use sturdy, easy-to-clean buckets or troughs. Position them in shaded areas to keep water cool and encourage drinking during hot weather.

- <u>Winter Water Solutions</u>: In colder climates, use heated water buckets or submersible heaters to prevent water from freezing. Check water sources frequently in winter to ensure availability.

- <u>Contamination Prevention</u>: Keep water containers elevated or secure to prevent contamination from bedding, manure, or curious goats climbing into them.

Daily Farm Management:
Routine Care for Healthy, Happy Goats

Establishing a consistent daily routine will keep your goats healthy and make your farm operations more efficient. Goats thrive on routine, and following a predictable schedule can reduce their stress and make care tasks easier.

Daily Routines to Follow

- <u>Feeding</u>: Feed your goats at the same time each day. Provide hay, minerals, and any necessary supplements. During winter or times of pasture scarcity, add grain or concentrate to their diet.

- <u>Water Checks</u>: Ensure water containers are clean and full. In summer, replace water regularly to keep it fresh.

- <u>Health Checks</u>: Perform quick daily health assessments by observing your goats. Look for signs of illness, such as lethargy, lack of appetite, or unusual behavior.

- <u>Milking (if applicable)</u>: If you have dairy goats, set up a consistent milking schedule. Goats prefer routine, and regular milking keeps them comfortable and maintains milk production.

Safety and Security on the Farm

Running a goat-friendly homestead involves more than just providing food and shelter. You must also take steps to protect your goats from harm, whether it's predators, toxic plants, or farm hazards.

Additional Safety Measures

- <u>Predator Protection</u>: If your area is prone to predators like coyotes, foxes, or stray dogs, consider adding a livestock guardian animal (such as a dog or donkey) or installing electric fencing. Secure housing at night is essential.

- <u>Toxic Plants and Chemicals</u>: Goats are curious eaters and may consume toxic plants if given the chance. Familiarize yourself with local poisonous plants and remove them from the pasture or fence them off. Store chemicals and feed in secure areas.

- <u>Food Storage Safety</u>: Keep feed bins sealed to prevent contamination from rodents or access by goats. Goats are notorious for getting into storage bins and overindulging, which can lead to serious health problems like bloat.

Organizing for Maximum Efficiency

Running an efficient homestead requires organization and planning. Here are some tips to streamline your operations:

- <u>Designate Specific Areas</u>: Assign different areas for feeding, milking, and health checks to create a flow in your daily routine. This makes it easier to manage multiple tasks and reduces confusion among the goats.

- <u>Record Keeping</u>: Maintain records of vaccinations, breeding schedules, and health checks. A simple notebook or digital spreadsheet can help you stay on top of your herd's care.

- <u>Emergency Supplies</u>: Keep a well-stocked first aid kit specifically for goats. Include items such as antiseptic sprays, bandages, hoof trimmers, dewormers, and electrolyte solutions.

Conclusion

Setting up your farm or homestead for goats is a rewarding process that lays the foundation for a successful, stress-free goat-keeping experience. By building secure shelters, installing reliable fencing, managing pastures effectively, and following a consistent care routine, you ensure that your goats will thrive. Thoughtful planning also minimizes risks and helps you manage your time and resources efficiently.

With the right setup, your goats will have a safe and comfortable home, and you'll enjoy the peace of mind that comes with knowing they are well cared for. The next chapter will dive deeper into feeding and nutrition—another essential aspect of maintaining a healthy, productive herd.

Chapter Four:
Feeding and Nutrition

Feeding goats properly is at the heart of good animal husbandry. Nutrition plays a crucial role in their health, growth, milk production, and overall well-being. A goat's diet must meet their specific needs based on age, purpose (milk, meat, or fiber production), and environmental factors like climate and seasons. While goats are known for being indiscriminate eaters, they are actually selective and have specific nutritional requirements that, if not met, can lead to health problems. This chapter will guide you through the essentials of a balanced diet, the importance of supplements, seasonal feeding strategies, and how to adapt their nutrition to different life stages. Understanding these fundamentals will ensure your goats remain healthy, productive, and content on your homestead.

The Foundation of a Goat's Diet:
Forage, Hay, and Shrubs

At the core of a goat's diet is forage, which includes hay, shrubs, and pasture grasses. As ruminants, goats rely on fiber-rich food to support their four-chambered stomachs, particularly the rumen, which ferments fibrous materials and converts them into energy. A diet that lacks sufficient fiber can lead to serious digestive issues, such as bloat or ruminal acidosis. Providing high-quality hay and access to shrubs ensures your goats get the nutrition they need year-round, especially during seasons when fresh pasture isn't available.

Hay serves as the primary feed, particularly in winters when pastures lie dormant. Two main types of hay are commonly fed to goats: alfalfa and grass hay. Alfalfa is a legume hay that is high in protein and calcium, making it especially beneficial for lactating and growing kids. However, it can pose risks for male goats, such as urinary calculi (stones), due to its high calcium content. Grass hay varieties, including Timothy, Bermuda, and orchard grass, are lower in protein and better suited for adult bucks, wethers, and non-lactating goats. Providing hay free-choice—so goats can eat as much as they need—supports healthy digestion and keeps their rumen functioning properly.

In addition to hay, goats love to feed on shrubs, weeds, and tree leaves. These not only provide essential nutrients but also stimulate natural behaviors, reducing boredom. Goats prefer variety in their diet, and offering them different types of forage helps keep them healthy and engaged. Common food items include blackberry bushes, mulberry leaves, and certain types of weeds. However, it's crucial to ensure the plants they consume are non-toxic, as goats can sometimes nibble on harmful vegetation.

Essential Supplements: Minerals and Vitamins for Goat Health

While hay and shoots or leaves form the foundation of a goat's diet, goats require supplemental minerals and vitamins to maintain optimal health. Copper, selenium, and calcium are among the most important minerals, each serving a specific function in a goat's body. Copper deficiency is a common issue, particularly in areas where the soil is naturally low in copper. Symptoms include a dull or faded coat, weight loss, and reproductive problems. However, care must be taken not to over-supplement, as copper toxicity can be just as dangerous. Selenium supports muscle development and reproductive health, but goats in selenium-deficient regions may require additional supplementation to prevent issues like white muscle disease, a condition that affects the muscles of young kids. Calcium is essential for lactating to prevent milk fever (hypocalcemia), a condition that can arise from low blood calcium levels.

To ensure goats receive the necessary minerals, it's best to provide them with free-choice loose minerals formulated specifically for goats. Avoid using mineral blocks or mixes meant for sheep, as these lack sufficient copper. Place the minerals in sheltered areas to keep them dry and accessible year-round. Additionally, offering goats baking soda free-choice can help balance stomach acidity, particularly if they are prone to digestive upset.

Grains and Concentrates: When and How to Use Them

Grains and concentrates are energy-dense feeds that can be beneficial for certain goats, but they must be fed in moderation. While high-production dairy goats, growing kids, and meat goats may require grain to meet their energy needs, overfeeding can lead to enterotoxemia (overeating disease) or bloat, both of which are life-threatening. Grain should never be the primary food source for goats—it should only complement their forage intake.

When feeding grain, it's important to introduce it gradually. Sudden changes in diet can disrupt the rumen and cause digestive problems. Dairy goats, for example, may benefit from grain supplements during peak lactation to maintain body condition and support milk production. Meat goats may receive additional grain to promote muscle development, especially before market or slaughter. Kids can begin eating small amounts of grain once their rumen develops, typically at around three to four weeks of age. However, grain should be avoided for adult wethers and bucks unless they are underweight, as it can increase the risk of urinary calculi.

A general rule of thumb is to limit grain intake to half a pound per goat per day and adjust based on individual needs. Monitor their body condition regularly and reduce grain if they gain excess weight. Always ensure that grain is fed in combination with hay to keep the rumen functioning properly.

Nutritional Needs Across Different Life Stages

The nutritional needs of goats vary throughout their lives, and it's essential to adjust their diet accordingly. Kids, for instance, require more protein and energy to support rapid growth. Access to high-quality hay, small amounts of grain, and free-choice minerals helps them develop strong muscles and bones. Pregnant does have increased energy demands, especially in the final month of gestation when most fetal growth occurs. Feeding them alfalfa hay, along with a small amount of grain, prevents pregnancy toxemia, a metabolic disorder caused by insufficient energy intake.

Lactating does have the highest nutritional requirements, as their bodies need extra energy, protein, and calcium to produce milk. Alfalfa hay, which is rich in both protein and calcium, is ideal for dairy goats during this period. Additionally, feeding grain during milking helps maintain their body condition and supports consistent milk production. Aging goats may face dental issues that make chewing hay difficult, so switching to softer feed options—such as soaked alfalfa pellets—ensures they continue to receive adequate nutrition without discomfort.

Water:
The Most Essential Nutrient

Water is the most overlooked yet essential component of a goat's diet. Goats need constant access to clean, fresh water to stay healthy. Dehydration can lead to decreased feed intake, digestive and urinary problems, particularly in bucks and wethers. On average, a goat drinks 1-3 gallons of water daily, though this amount increases during hot weather and lactation. Lactating goats require significantly more water to support milk production.

Maintaining water quality is crucial, as goats are finicky drinkers who may refuse stale or dirty water. In cold climates, preventing water from freezing is a challenge. Heated water buckets or submersible heaters ensure goats can stay hydrated during the winter months. Position water containers in shaded areas during summer to keep the water cool and appealing. Regularly cleaning and refilling water troughs prevents contamination and encourages healthy drinking habits.

Seasonal Feeding Strategies

Goat feeding strategies should adapt to the changing seasons to meet their nutritional needs. In spring and summer, goats can rely more on pasture forage, reducing the need for supplemental hay. However, mineral supplements should still be offered year-round to prevent deficiencies. In fall and winter, when pasture quality declines, hay becomes the primary feed. Increasing hay portions during the colder months helps goats maintain body heat, as they burn more calories to stay warm. Offering tree branches or evergreen boughs during winter provides additional fiber and mimics their natural browsing behavior.

Supplementing with grain during winter can provide the extra energy goats need, especially for goats exposed to cold temperatures. However, it's essential to monitor their weight and adjust portions to prevent overfeeding, as reduced activity levels in winter can lead to weight gain.

Free-Choice Mineral Feeding and Monitoring

Providing free-choice minerals ensures that goats have access to essential nutrients whenever they need them. Loose minerals are generally more effective than blocks, as goats can consume them in smaller, more frequent doses. Place mineral feeders in dry, sheltered locations to keep them accessible in all weather conditions. Monitor your goats' mineral consumption regularly—some goats may consume more minerals than others, which could indicate an underlying nutritional imbalance. Detailed records of mineral use, feeding schedules, and body condition can provide valuable insights into your herd's nutritional needs and help you make informed adjustments over time.

Conclusion

Feeding goats is both a science and an art, requiring observation, planning, and flexibility. Understanding the role of forage, minerals, and grain in your goats' diet ensures they receive the right nutrition for every life stage. Seasonal feeding strategies and consistent access to clean water further support their health and productivity. By staying attentive to their individual needs and making thoughtful adjustments, you can maintain a happy, healthy herd year-round. In the next chapter, we'll explore goat health and medical care, providing practical advice on preventing common diseases, managing parasites, and ensuring your goats stay in top condition. With the right nutrition and care, your goats will flourish and become a cherished part of your homestead.

$\smile\!\!\infty$

Chapter Five:
Goat Health and Medical Care

Keeping goats healthy is one of the most vital aspects of successful goat husbandry. Goats are resilient animals, but they are still susceptible to various health issues that can significantly affect their quality of life and productivity. This chapter provides a comprehensive guide to maintaining goat health, addressing common diseases, preventive measures, and medical care protocols that every goat owner should know.

Routine Health Checks

Regular health checks are the foundation of good goat husbandry. These assessments help you monitor your goats' overall well-being, detect early signs of illness, and ensure they are thriving. Here's what to look for during routine health checks:

Body Condition

Assessing body condition is crucial for determining whether a goat is underweight, at an ideal weight, or overweight. Use a scale of 1 to 5, where:

- 1 indicates emaciation (very thin)

- 2 indicates thinness (visible ribs, bony structure)

- 3 indicates average condition (ribs are not visible but can be felt)

- 4 indicates fat (smooth appearance, no ribs visible)

- 5 indicates obesity (excess fat, difficulty moving)

Ideal body condition varies based on the goat's breed, age, and purpose (e.g., dairy, meat, fiber). Adjust feed and management practices based on body condition scores.

Coat Quality

The coat is a reliable indicator of a goat's health. A healthy goat typically has a shiny, smooth coat without bald patches or excessive scratching. If the coat appears dull, rough, or patchy, it may signal nutritional deficiencies, parasites, or underlying health issues.

Overall Well-Being

Check for general signs of health, including:

- Behavior: A healthy goat is alert, curious, and social. Changes in behavior, such as lethargy or isolation, can indicate illness.

- **Appetite**: Monitor food and water intake. A sudden decrease in appetite can be a warning sign of digestive issues or illness.

- **Eyes and Nose**: Bright, clear eyes and a clean nose are indicators of good health. Discharge or excessive tearing can be signs of respiratory infections or allergies.

Regularly performing these checks will help you catch potential health issues early, ensuring your goats remain in optimal condition.

Common Goat Diseases

Understanding common goat diseases is essential for prevention and treatment. Here are some prevalent issues to be aware of:

Internal Parasites (Worms)

Symptoms: Weight loss, poor coat condition, lethargy, diarrhea, and anemia (pale mucous membranes).

Causes: Goats are prone to internal parasites, particularly in moist, warm environments. Overcrowding and poor sanitation can exacerbate infestations.

Preventative Measures:

1. Implement a regular deworming schedule.

2. Rotate dewormers to avoid developing resistance.

3. Conduct fecal egg counts to monitor parasite levels and determine the need for treatment.

Respiratory Infections

Symptoms: Coughing, nasal discharge, labored breathing, and fever.

Causes: Respiratory infections can arise from viruses, bacteria, or environmental factors such as poor ventilation or exposure to drafts.

Preventative Measures:

1. Maintain good ventilation in barns.

2. Avoid overcrowding and wet bedding.

3. Monitor your herd for early signs of respiratory distress and isolate affected animals.

Coccidiosis

Symptoms: Diarrhea (often bloody), weight loss, dehydration, and poor appetite.

Causes: Coccidiosis is caused by protozoan parasites and often affects young goats under stress, such as weaning or transportation.

Preventative Measures:

1. Ensure clean, dry bedding to reduce oocyst buildup.

2. Provide proper nutrition and avoid overstocking.

3. Use anticoccidial medications as needed, particularly in high-risk situations.

Deworming Protocols

A regular deworming schedule is essential for maintaining goat health. Over time, parasites can develop resistance to dewormers, so it's important to rotate between different classes of dewormers. Here's how to do it:

1. Identify the Parasite: Conduct fecal egg counts to determine which parasites are present in your goats and guide your treatment approach. Fecal egg counts should ideally be conducted at a veterinary clinic, especially for beginners, as they require specialized equipment and expertise to accurately identify parasites. Many veterinarians offer this service, and regular testing helps you avoid unnecessary treatments and manage parasite load effectively.

2. Choose Dewormers: Based on the results of the fecal egg count or the vet's recommendation, select the appropriate dewormer. It's essential to rotate between different classes of dewormers to prevent resistance. Common classes include:

   - Benzimidazoles (e.g., fenbendazole)

   - Imidazothiazoles (e.g., levamisole)

   - Macrocyclic lactones (e.g., ivermectin)

3. Administration: Dewormers can be administered orally or via injection, depending on the product used. Follow dosage recommendations based on the goat's weight.

4. Monitor Effectiveness: After treatment, conduct another fecal egg count to ensure the dewormer was effective. If high

egg counts persist, consult a veterinarian for alternative treatments.

Vaccination Protocols

Vaccinations are crucial for preventing common diseases in goats. Here are the recommended vaccines and their administration schedules:

CD&T Vaccine

The CD&T vaccine protects against Clostridium perfringens type C and D and Tetanus. It is essential for all goats.

- Initial Vaccination: Kids should receive their first shot at 6-8 weeks.

- Booster: A booster shot is given at 10-12 weeks.

- Annual Booster: All goats should receive an annual booster thereafter.

Other Vaccines

Depending on your area and specific risks, consider other vaccines such as those for:

- Caseous Lymphadenitis (CL)

- Foot and Mouth Disease (FMD)

- Leptospirosis

Consult with a veterinarian to tailor a vaccination schedule based on your herd's specific needs and local health concerns.

Rabies Vaccine

Rabies is a fatal disease that can affect all mammals, including goats.

Note that in Canada, rabies vaccination is not widely recommended for goats. This vaccine is typically limited to animals like dogs, cats, ferrets, and certain horses, as it must be administered by a veterinarian.

<u>Administration</u>: Typically given at 12 weeks of age or older, with a booster recommended annually.

VCPR (Veterinarian-Client-Patient Relationship)

In Canada, a valid Veterinarian-Client-Patient Relationship (VCPR) must be established to obtain antibiotics, certain dewormers, and other prescription products. This relationship ensures that a licensed veterinarian is familiar with your herd and can prescribe treatments as needed. A valid VCPR includes regular veterinary check-ups and consultations, which allow the veterinarian to make informed decisions about your herd's health.

Controlling External Parasites

External parasites, such as lice, mites, and ticks, can cause significant discomfort and health issues in goats. Here are strategies for managing these pests:

Identification

Regularly inspect your goats for signs of external parasites, including excessive scratching, hair loss, and visible insects. Lice are often detected as tiny, moving specks on the skin, while ticks may attach to the skin.

Control Measures

- <u>Maintain Clean Bedding</u>: Regularly clean and replace bedding to minimize parasite populations.

- <u>Dusting and Spraying</u>: Use appropriate insecticides specifically designed for goats. Apply dusts or sprays according to label instructions, paying attention to safety guidelines.

- <u>Rotate Pasture</u>: Rotating goats to different pastures can help reduce parasite loads in specific areas.

Good Hygiene Practices

Good hygiene is essential for parasite control. Clean water, proper nutrition, and adequate space contribute to overall health, making goats less susceptible to infestations. Regular grooming can also help detect and manage external parasites early.

Hoof Care

Proper hoof care is vital for goat health, as neglected hooves can lead to serious issues such as foot rot and abscesses. Here's how to maintain healthy hooves:

Regular Trimming

Goat hooves should be trimmed every 6-8 weeks, depending on the goat's living conditions and activity levels. Here's how to trim hooves effectively:

1. Gather Equipment: You'll need hoof trimmers, a hoof knife, and possibly a rasp for smoothing edges.

2. Restrain the Goat: Secure the goat in a comfortable position. Some owners prefer to have an assistant to help hold the goat steady.

3. Examine the Hooves: Inspect the hoof for cracks, splits, or signs of infection.

4. Trim the Hooves: Trim the outer edges of the hooves, taking care not to cut into the quick (the sensitive part inside the hoof). Remove excess hoof material and shape the hoof to prevent cracking.

5. Treating Hoof Problems: For foot rot or abscesses, consult a veterinarian for proper treatment. Footrot requires diligent cleaning and may necessitate antibiotic treatment.

Monitoring Hoof Health

Regular hoof care not only promotes healthy feet but also allows for early detection of issues. Always watch for changes in gait or signs of discomfort.

Emergency First Aid

Accidents and emergencies can happen, and being prepared is key to effectively managing such situations. Here's a guide to emergency first aid for goats:

Minor Injuries

For minor injuries like cuts and abrasions:

1. Clean the Wound: Use warm, soapy water or a saline solution to gently clean the area.

2. Apply an Antiseptic: Use a veterinary-approved antiseptic spray or ointment to prevent infection.

3. Bandage if Necessary: If the injury is deep or bleeding, consider bandaging it to keep it clean.

Administering Medications

If you need to give medications, ensure you know the correct dosage and administration route. Always read the label and consult with a veterinarian if you have questions.

Critical Situations

Bloat

Bloat is a serious condition where gas builds up in the rumen, causing discomfort and potentially life-threatening situations. Signs include a distended abdomen and signs of distress.

- Treat Immediately: Administer an anti-bloat medication or mineral oil if available.

- Seek Veterinary Help: If the goat does not improve quickly, contact a veterinarian.

Poisoning

If you suspect poisoning, determine the source if possible and contact a veterinarian immediately. Common sources of poisoning include toxic plants, chemicals, and certain human foods.

Well-Stocked First Aid Kit

Prepare a first aid kit tailored to your goats' needs. Essential items include:

- Antiseptic wipes and sprays

- Bandages and gauze

- Hoof trimmers and hoof care supplies

- Thermometer (for monitoring temperature)

- A list of emergency contacts, including your veterinarian

Regularly check and replenish your first aid kit to ensure it's always ready.

Maintaining the health and well-being of your goats is a rewarding but responsible endeavor. By conducting routine health checks, understanding common diseases, implementing preventative measures, and being prepared for emergencies, you can ensure your goats lead healthy, productive lives on your homestead.

$\smile\!\!\propto$

Chapter Six:
Breeding, Kidding, and Raising Kids

Raising goats from birth is one of the most rewarding experiences for any goat keeper. From selecting the right breeding stock to witnessing the first steps of a newborn kid, breeding and kidding add a deeper dimension to goat care. A well-planned breeding program allows you to improve the genetics of your herd, select desirable traits, and expand your operation sustainably. This chapter will guide you through the entire process of breeding, kidding, and raising kids, including selecting the best breeding stock, understanding the reproductive cycle, managing breeding safely, and caring for both does and kids through each stage.

Basics of Goat Breeding

Breeding begins with the selection of suitable stock. When choosing goats to breed, it's important to prioritize health, temperament, and productivity traits, such as milk yield for dairy breeds, muscle tone for meat goats, or fleece quality for fiber goats. Additionally, both the doe and the buck should be free from genetic health issues, such as malformations or susceptibility to diseases. A healthy doe and buck will produce stronger offspring, better suited to thrive and contribute positively to your herd.

The goat reproductive cycle, or estrous cycle, occurs roughly every 18-21 days, lasting about two to three days within that period. During this time, does experience estrus, or "heat," which is when they are receptive to mating. Signs of heat in a doe may include increased vocalization, a swollen or reddened vulva, tail wagging, a decrease in milk production for dairy goats, and unusual affection toward other goats. Recognizing these signs is key to timing the mating process successfully, especially if you're planning for a specific kidding season.

Managing the Breeding Process: Timing and Mating

The ideal time to breed a doe is once she is at least eight months old or has reached 60-70% of her expected adult weight, which ensures she's physically developed enough to carry and birth kids safely. Some dairy breeds can require a bit longer to mature. Bucks are generally fertile by five to six months, though it's common to wait until they're closer to one year to ensure full physical maturity. When introducing a buck to a doe, it's often best to place them together in a neutral area where neither has a strong territorial attachment. Bucks can be quite eager, so make sure you supervise the initial introduction to prevent stress on the doe.

If you're breeding multiple does, creating a breeding management sheet is helpful for keeping track of important dates and details. This sheet should include each doe's name, age, weight, date of mating, and expected kidding date, as well as notes on the breeding behavior observed. By keeping thorough records, you'll be better prepared for kidding season and can monitor each doe's health and progress throughout gestation.

Pregnancy:
Monitoring Health and Signs of Labor

A doe's pregnancy, or gestation period, typically lasts between 145 to 155 days, with an average of 150 days. Throughout pregnancy, it's important to monitor the doe's health, as well as to ensure she receives proper nutrition and care. A balanced diet high in protein and calcium supports fetal growth and prepares the doe for milk production. Be cautious not to overfeed, as excess weight gain can make labor difficult. Regular health checks are essential, especially in the last month, to monitor for signs of pregnancy toxemia (a metabolic disorder resulting from energy deficiency) and other potential complications.

In the final week of pregnancy, signs of impending labor may include a drop in the doe's abdomen as the kids settle into the birthing position, swelling of the udder (often called "bagging up"), and relaxation of the ligaments around the tail head, causing a visible hollow on either side of her tail. You may also notice a change in the doe's behavior—she might become more restless, paw at the ground, or separate herself from the herd as she prepares to give birth.

Preparing for Kidding:
Setting Up a Safe Kidding Area

As the doe nears her due date, it's time to prepare a clean, quiet, and safe kidding area. This area should be sheltered, dry, and equipped with fresh bedding to provide comfort and reduce the risk of infections. A secluded kidding pen also gives the doe privacy and a calm environment, which can reduce stress during labor. Essential supplies for the kidding area include clean towels, palpation gloves, iodine, and a flashlight or lantern for visibility if kidding occurs at night.

Additionally, having a fully stocked kidding kit nearby is crucial. This kit ensures that you are prepared for any situation that may arise during labor and delivery, helping you provide immediate assistance if necessary. Here are the recommended items to include in a well-prepared kidding kit:

- Clean Towels: For drying off newborn kids to prevent hypothermia and stimulate circulation.

- Bulb Syringe: Useful for clearing mucus from the kids' nose and mouth to ensure clear airways. It can also be used alongside a straw and stuck up the nose quickly to stimulate breathing.

- Thermometer: A rectal thermometer to monitor the temperature of both the doe and her kids, especially if there are concerns about hypothermia.

- Disposable Palpation Gloves: Essential for maintaining hygiene and preventing infections when handling the doe or assisting during delivery.

- <u>Sterile Scissors</u>: For cutting the umbilical cord if necessary, which should be done with sterile tools to prevent contamination.

- <u>Iodine Solution</u>: A 7% iodine solution for adding to the water used in assisting with delivery.

- <u>Lubricant</u>: In case you need to assist with delivery, lubricant can help reduce friction and make the process smoother and less stressful for the doe.

- <u>Electrolyte Solution</u>: A powdered or liquid electrolyte solution for rehydrating the doe if she becomes weak or exhausted during labor.

- <u>Betadine or Antiseptic Solution</u>: For cleaning any minor wounds or scrapes that may occur during delivery.

- <u>Heating Pad or Heat Lamp</u>: For keeping kids warm if they are born in colder temperatures or appear weak and chilled.

- <u>Notebook and Pen</u>: To record essential details such as the time of birth, number of kids, their sex, and any observations about the doe's condition.

- <u>Sugar or Molasses</u>: This can be added to warm water and offered to the doe after kidding to provide a quick source of energy.

- <u>Colostrum:</u> To feed the kid or kids if the mother is unable to do so.

- <u>Goat Tuber:</u> To tube colostrum for the kids.

- <u>Goat Kid Bottle and Nipple:</u> Useful for bottle-feeding kids.

With these items on hand, you'll be well-prepared to handle a wide range of potential situations, helping ensure a smooth delivery process for the doe and a healthy start for her kids. A properly stocked kidding kit offers peace of mind and the ability to provide immediate care when it matters most.

The Birthing Process

Labor in goats occurs in three stages: dilation, active labor, and delivery of the placenta. In the first stage, the doe's cervix dilates, and she may appear restless, paw at the bedding, or get up and down frequently. This stage can last several hours, and while it may seem prolonged, it's important to remain patient as the doe prepares for active labor.

During active labor, contractions become stronger, and the doe will begin pushing. At this point, you may notice the amniotic sac appear as a bubble at the vulva, which indicates that birth is imminent. Kids are typically born in a nose-and-feet-first position. After the first kid is delivered, the doe will usually rest briefly before delivering any additional kids. Multiple births are common, with twins or triplets being the norm in many breeds.

While most goats deliver kids without issues, it's important to be prepared to assist if complications arise. If labor stalls or the doe is pushing without progress for more than 30 minutes, you may need to gently assist. Clean your hands thoroughly and use lubrication if you must enter the birth canal. If the kid is in an unusual position—such as breech (rear end first) or with its head turned back—attempt to reposition it carefully. However, it's always best to consult a veterinarian if complications persist, as timely intervention can prevent serious outcomes for both the doe and her kids.

Birthing Malpresentations:
How to Correct Them

The correct birthing position for a kid is with the head and two front feet, feet pointing downward, and the head resting on the legs. Kids can be born in different positions, but these positions often pose a higher risk and may require assistance. It's important to monitor and assist if needed. The key is not to rush but also not to wait too long. Some can still successfully birth kids in misrepresented positions, depending on the doe and the kid's size.

TIP 1: When attempting to reposition a kid, protect the uterus at all times. Use your hand to cover any feet, head, or teeth to avoid puncturing the uterine wall.

TIP 2: Before repositioning, identify the position of the kid by gently feeling around to find "landmarks," such as the hooves. Pay attention to their orientation—whether the hooves point up or down—and determine if you feel a head or tail and whether they're positioned upward or downward. This will guide you in repositioning.

- Head Back or Down: Gently locate the legs and follow them to the neck and head. Try to gently grasp the neck and head, carefully bringing it toward the birth canal. Always cover the teeth with your hand to protect the uterus.

- Head First with No Legs: Locate the head and follow along the neck to the shoulders and legs. Gently grasp a hoof and pull it toward the canal. If the birth has not progressed too far, you may be able to gently push the kid's head back into the canal to get the legs into position. However, if the head has emerged too far, you won't be able to push it back. Instead, insert your hand under the head and neck to lift the shoulders slightly, but avoid pulling the neck too hard to prevent injury to a live kid. Prolonged head-first positioning without legs may be fatal to the kid.

- <u>Leg(s) Down or Back</u>: Locate the head and neck, trace down to the legs, and gently bring them up into the canal.

- <u>Upside Down</u>: Some kids can be born upside down with assistance, but if possible, try to rotate them into the upright position before pulling.

- <u>Sideways</u>: Rotate the kids slightly to get them upright and in the correct birthing position.

- <u>Backward (Back Legs/Tail First)</u>: When the kid presents back legs first, ensure both back legs are in the canal before assisting in delivery. Be aware that, in this position, the umbilical cord is compressed against the pelvis, cutting off oxygen. It's essential to deliver the kid quickly to minimize the risk of oxygen deprivation.

- <u>Breech (Rear End First)</u>: In a breech position, gently push the kid forward and down to maneuver the back legs up into the canal, allowing delivery in a backward position.

- <u>The BIG Kid</u>: Sometimes, kids are simply too large to pass through the birth canal, a common issue with single births. If the head and feet cannot be repositioned into the canal, veterinary assistance is required, and a C-section may be necessary.

- <u>Ringwomb (Cervix Doesn't Dilate Properly)</u>: Occasionally, a doe's cervix does not fully dilate during labor, a condition known as ringwomb. This occurs more commonly in first-time cases but can happen at any age. If the doe is in active labor without full dilation, she may have ringwomb, which requires immediate veterinary intervention. In most cases, a C-section will be necessary.

Recognizing and addressing these malpresentations with care will help support a safe birthing process for both the doe and her kids.

Post-Kidding Care: Ensuring Health for Doe and Kids

After delivery, the doe may rest, but it's important to monitor both her and the kids closely. The first critical step for the kids is to receive colostrum, the nutrient-rich milk that provides essential antibodies and energy. Colostrum should ideally be consumed within the first hour of life, as a kid's ability to absorb these antibodies diminishes rapidly within the first 24 hours. If the doe's colostrum supply is inadequate, you may need to provide a colostrum substitute.

Inspect the kids for signs of weakness or difficulty breathing. Use a clean towel to dry them off, which helps prevent hypothermia and stimulates the kids. Ensure that each kid is nursing well or, if necessary, assist by guiding them to the test. Bottle-feeding may be necessary for kids who struggle to nurse or are weak, as well as for those who are weak, to ensure they receive adequate nutrition.

Check the doe for retained placenta and any signs of distress, as a healthy postpartum recovery is essential for her return to the herd. She should pass the placenta within 12 hours, and if not, consult a veterinarian to avoid complications like infection. If the placenta isn't passed within 12 hours, antibiotics are usually given. Offering her fresh water and an electrolyte solution can help her recover and make sure she has access to high-quality hay to support her milk production.

Raising Kids:
Feeding, Health Checks, and Socialization

Once the initial post-kidding phase has passed, the focus shifts to raising healthy kids. Nursing kids will receive all the necessary nutrients from their mother's milk for the first few weeks, but some homesteaders opt to bottle-feed, especially in cases of weak kids or when the doe cannot produce enough milk. If you choose to bottle-feed, use a milk replacer specifically formulated for goats and follow a consistent schedule of four to six feedings per day, gradually reducing frequency as the kids grow.

Around two to four weeks, kids will begin nibbling on solid foods, which helps stimulate rumen development. At this stage, provide high-quality hay, clean water, and a small amount of kid-safe grain or starter pellets to support their growth. Free-choice minerals designed for young goats should also be accessible to meet their nutritional needs.

Routine health checks are essential for early disease detection and development tracking. Inspect kids regularly for any signs of illness, such as nasal discharge, diarrhea, or lethargy, and check that their hooves are growing correctly. Kids should receive vaccinations, including the CD&T vaccine, at approximately six to eight weeks, with a booster four weeks later, to protect them from clostridial diseases and tetanus.

Socializing kids from a young age fosters strong bonds with humans and the rest of the herd. Handle them gently but frequently, especially if they will be part of a dairy operation, as this familiarizes them with human contact and makes future milking or health checks easier. Gradually introduce them to other herd members, allowing supervised interaction to avoid injury from older goats. Kids are naturally playful and active, so creating a safe, enriched environment with opportunities to climb, explore, and play supports their physical and mental development.

Conclusion

Breeding, kidding, and raising kids are pivotal experiences that allow goat keepers to witness the full life cycle of their herd. From selecting breeding stock to guiding kids through their early days, each stage requires careful planning, observation, and adaptability. A well-executed breeding program not only adds value to your herd but also deepens your understanding of goats and strengthens your connection to them. By following these guidelines and preparing for the unexpected, you'll be well-equipped to support your does through pregnancy and ensure your kids start life healthy and strong.

Chapter Seven:
Milking and Dairy Goat Care

Milking dairy goats is a rewarding experience that allows you to produce fresh, nutritious milk for your household and even create dairy products like cheese and yogurt. To ensure both high-quality milk and the health of your dairy goats, it's essential to establish a consistent milking routine, practice good hygiene, and prioritize the overall care of the goats. This chapter covers everything you need to know about milking goats, from selecting the right dairy breeds to maintaining udder health, storing milk, and setting up a reliable routine. Whether you are a new goat keeper or experienced in dairy goat care, the following sections will provide a comprehensive guide to help you make the most of your dairy herd.

Choosing the Right Dairy Goat and Establishing a Milking Routine

Selecting the right dairy goat breed is the foundation of successful milking. Dairy goats are bred specifically for milk production, and popular breeds include Nubians, Saanens, Alpines, and LaManchas. Each of these breeds offers unique characteristics—Nubians, for example, are known for their high butterfat content, which is ideal for cheese making, while Saanens are known for their high milk yield. Choose a breed based on your production goals and personal preferences, considering factors like milk quantity, butterfat content, and temperament.

Once you have selected your goats, establish a milking routine that is consistent and stress-free for both you and the goats. Milking should be done at the same times each day, typically every 12 hours, to ensure a steady milk supply. A typical schedule might involve milking early in the morning and again in the evening. Goats are creatures of habit, and a reliable routine helps them feel secure and comfortable, positively affecting milk production. Setting up a dedicated milking area also helps streamline the process; this area should be clean, well-ventilated, and free from distractions to create a calm environment for the goats.

Milking Techniques: Ensuring Health and Quality

Proper milking techniques are critical for maintaining the health of your goats and the quality of the milk. Begin by cleaning the udder with a warm, damp cloth to remove dirt and bacteria. Using an udder wash or disinfectant is recommended to prevent contamination. Massaging the udder gently before milking helps stimulate milk letdown, making the process smoother. The hand-milking method involves gently squeezing the teat between the thumb and forefinger at the top and gradually releasing pressure down to the tip. Avoid pulling down on the teats, as this can cause discomfort or damage to the udder. Instead, use a gentle, rhythmic squeezing motion to express milk efficiently.

Hygiene is paramount in milking. Always wash your hands thoroughly before starting, and ensure all equipment, including buckets, strainers, and jars, is sterilized before each use. Stainless steel or glass containers are ideal for collecting milk, as they are easier to sterilize and don't absorb odors or bacteria. Additionally, avoid using any equipment that has been cracked or damaged, as these imperfections can harbor bacteria. Keeping your milking environment, hands, and equipment clean not only ensures high-quality milk but also helps protect your goats from infections like mastitis.

Maintaining and Boosting Milk Production

To support consistent milk production, dairy goats require a well-balanced diet rich in protein, calcium, and vitamins. High-quality forage, such as alfalfa hay, is an excellent source of calcium and protein, which are essential for milk production. Supplementing with grains provides additional energy but should be done carefully, as overfeeding can lead to health issues. Lactating does benefit from feed specifically formulated for dairy goats, which includes the necessary vitamins and minerals to support milk production. Make sure they always have access to free-choice minerals, particularly calcium, which helps prevent milk fever, a condition caused by low calcium levels.

Hydration is equally important, as a lactating doe needs plenty of water to produce milk. A single dairy goat may drink up to 2-3 gallons of water per day, especially during warm weather. Ensure they have access to clean, fresh water at all times, as dehydration can lead to a decrease in milk supply. Stress reduction also plays a role in milk production. Goats are sensitive animals, and changes in their environment, loud noises, or unfamiliar people can affect their production. A consistent, calm routine, along with minimal disruptions, during milking time can help maximize yield. Monitor each goat's body condition and milk yield regularly, adjusting feed and care as needed to support their individual production levels.

Udder Health:
Preventing and Treating Mastitis

Keeping the udder healthy is essential for a productive dairy goat. Mastitis, an infection of the mammary gland, is one of the most common health issues in dairy goats and can significantly impact milk production. Symptoms of mastitis include swelling, heat, pain, or hardness in the udder, as well as changes in the milk's appearance, such as clots or discoloration. Early detection is key to successful treatment. Regularly examine each doe's udder for signs of inflammation and monitor the milk's consistency. If you suspect mastitis, consult a veterinarian for a proper diagnosis and treatment plan, which may include antibiotics or anti-inflammatory medication.

To help prevent mastitis, ensure that the udder and teats are thoroughly cleaned before and after each milking session. After milking, applying a teat dip or antiseptic solution helps kill any bacteria left on the teats. It's also important to dry the udder completely before allowing the goat to leave the milking area, as moisture can encourage bacterial growth. Maintain a clean milking area and bedding to minimize exposure to environmental pathogens and avoid abrupt changes in milking routines, as inconsistency can lead to stress and susceptibility to infections. Practicing good milking hygiene and monitoring udder health are the best defenses against mastitis.

Storing and Processing Goat Milk

Once milk is collected, proper storage and processing are crucial to maintain its freshness and quality. Goat milk spoils more quickly than cow's milk, so refrigeration should be done immediately after milking. Strain the milk through a fine, food-grade filter or cheesecloth into a clean container to remove any debris or contaminants. Store the milk in airtight glass jars or stainless-steel containers, as plastic can absorb flavors and odors, affecting the taste of the milk. Place the jars in the coldest part of the refrigerator and maintain a temperature of 35-40°F (1-4°C) for optimal freshness. Under these conditions, raw goat milk typically lasts about 5-7 days.

If you plan to pasteurize the milk, heat it to 165°F (74°C) for 15 seconds or to 145°F (63°C) for 30 minutes, then cool it rapidly by placing the container in an ice bath. I use my instant pot on the sterilized setting because it holds the temperature for the needed amount of time. Pasteurization kills harmful bacteria while preserving the milk's nutritional content. Pasteurized milk will last longer in the refrigerator—typically 7-10 days—compared to raw milk. For longer-term storage, milk can be frozen, although it may separate when thawed. Freezing in small batches and stirring after thawing helps restore a more even texture. Avoid refreezing thawed milk, as this can degrade its quality.

For those interested in home dairy processing, goat milk can be transformed into a variety of products, including cheese, yogurt, and soap. Cheese making requires specific tools, such as molds, cultures, and a cheese press, and is a great way to utilize surplus milk. Yogurt requires a starter culture and careful temperature control, while goat milk soapmaking involves mixing with oils and lye. Experimenting with these processes can add value to your dairy operation and provide unique, homemade products for your family or market.

Essential Milking Equipment
and Tools for Home Dairy Production

Setting up a home dairy requires some essential equipment to ensure safe and efficient milk handling. Milking stands are an invaluable tool, providing a secure, elevated platform for milking and helping the goat stay calm and still. Stands typically include a head catch and are adjustable for different sizes, making milking easier and more comfortable for both you and the goat. Stainless steel buckets or pails are ideal for collecting milk, as they are easy to sterilize and resistant to bacterial buildup. Milk strainers with fine mesh or disposable filters ensure that milk is free from impurities before storage.

For storing milk, glass jars or stainless-steel containers are recommended, as they do not absorb odors and maintain the milk's freshness. If you're pasteurizing milk at home, a thermometer is essential for monitoring temperatures accurately. For those making dairy products, additional equipment like cheese molds, presses, and yogurt makers can be beneficial. Storage racks or dedicated shelves in the refrigerator or freezer help keep dairy products organized and properly stored. Proper equipment not only simplifies the milking and processing routine but also ensures a higher standard of hygiene and quality for your dairy products.

Conclusion

Milking and caring for dairy goats is a fulfilling endeavor that allows you to produce fresh milk and explore a range of dairy products. By establishing a consistent milking routine, practicing good hygiene, and prioritizing the health of the goats, you'll create a smooth and productive dairy operation. Regularly monitoring udder health and using clean, high-quality equipment help ensure that your goats stay healthy and that the milk you collect is safe and nutritious. Goat milk storage and processing offer the chance to explore everything from raw milk to pasteurized options, along with specialty products like cheese and soap.

The journey of milking and dairy goat care is one of learning and adaptation, as each goat may have unique needs and preferences. As you gain experience, you'll refine your milking techniques, master milk-handling practices, and develop a deeper connection with your dairy goats.

Chapter Eight:
Things to Do with Goat Milk

Goat milk has been a cherished part of human diets for thousands of years, renowned for its distinctive qualities that differentiate it from other types of milk. In recent years, it has gained traction as a versatile and nutrient-rich alternative to cow milk, offering benefits that reach beyond the kitchen into skincare and sustainable farming practices.

Nutritional Benefits of Goat Milk

One of the standout features of goat milk is its remarkable digestibility. Unlike cow milk, goat milk contains smaller fat globules and lower levels of alpha-S1 casein, making it gentler on the digestive system for many people. This makes goat milk an excellent option for those with sensitive stomachs or mild lactose intolerance. Although not entirely lactose-free, goat milk contains less lactose than cow milk, which can help minimize discomfort for those who find traditional dairy products hard to digest.

Additionally, goat milk boasts an impressive nutrient profile. It is rich in essential vitamins and minerals, including calcium, phosphorus, magnesium, and potassium, all of which support bone health, muscle function, and overall wellness. Furthermore, goat milk has higher levels of medium-chain fatty acids than cow milk, which the body absorbs and converts into energy more efficiently.

Another key benefit is its natural richness in A2 casein, a type of protein linked to reduced inflammation and better gut health compared to the A1 casein prevalent in most cow milk. Research indicates that A2 casein is less likely to cause digestive issues, making goat milk a more appealing choice for individuals who experience discomfort after consuming dairy.

Overview of Goat Milk Versatility

Goat milk's versatility extends well beyond being a mere substitute for cow milk. Its creamy texture and mild flavor make it a favorite in culinary applications, ranging from artisanal cheeses and indulgent ice creams to tangy yogurts and rich kefir. Many chefs and home cooks prefer goat milk for its ability to enhance flavors while providing a delightful creaminess.

Beyond culinary uses, goat milk has carved out a niche in skincare. Its high levels of lactic acid, vitamins, and natural fats make it an exceptional ingredient in soaps, lotions, and various skincare products. Goat milk's gentle exfoliating properties and deep hydrating capabilities make it a popular choice for those seeking natural and effective skincare solutions.

Goat milk also presents significant entrepreneurial opportunities. Small-scale farmers and artisans can capitalize on its rising popularity by creating niche products, from gourmet cheeses to handcrafted soaps. With increasing consumer demand for sustainable and eco-friendly products, goat milk businesses can flourish by highlighting their commitment to ethical farming practices and high-quality production.

Sustainable and Eco-Friendly Aspects of Goat Farming

Raising goats is often more sustainable than keeping cows, particularly for small farms or households. Goats require less land, water, and feed compared to cows, making them ideal for small-scale farms or homesteaders. Additionally, goats are efficient grazers, thriving on a diverse range of vegetation, including plants that may not be suitable for other livestock.

Goat farming also aligns with regenerative agriculture practices, as goats help maintain healthy ecosystems by controlling overgrowth and enriching the soil with their manure. These benefits resonate with the growing focus on sustainability in food and farming, making goat milk an appealing choice.

Culinary Uses of Goat Milk

Goat milk presents an exciting array of culinary possibilities thanks to its creamy texture, robust flavor, and versatility in numerous recipes. Whether you're a home cook venturing into the world of artisanal cheeses or a dessert lover seeking something distinctive, goat milk provides endless opportunities to enhance your culinary creations. In the following sections, we will delve into its various applications, offering guidance on crafting goat milk-based delicacies along with tips for achieving success and inspiration for flavor experimentation.

Homemade Goat Cheese Recipes: Chevre

Chevre is a soft and tangy goat cheese that is not only delicious but also one of the simplest cheeses to make at home. With just a few ingredients and some patience, you can create a fresh cheese that can be enjoyed in various dishes or on its own. Here's a comprehensive guide to making your own chevre, including tips for success and flavor variations.

Ingredients and Tools Needed

- Fresh Goat Milk: 1 gallon

- Chevre Culture: Alternatively, you can use 1/4 cup of lemon juice

- Salt: To taste

- Tools: Cheesecloth, thermometer, colander, and a large pot

Step-by-Step Process

1. Heat the Goat Milk: Begin by slowly heating the goat milk in a large pot to 86°F (30°C). Stir gently to prevent scorching and ensure even heating.

2. Add Culture or Lemon Juice: Once the milk reaches the desired temperature, add the chevre culture or lemon juice. Mix thoroughly to ensure even distribution.

3. Let It Sit: Cover the pot and let the mixture sit at room temperature for 12-24 hours. During this time, the curds will form. The longer you let it sit, the tangier the cheese will become.

4. Drain the Curds: After the curds have formed, scoop them into a colander lined with cheesecloth. Allow the whey to drain out.

5. Hang the Cheese: Gather the corners of the cheesecloth and tie them into a bundle. Hang this bundle over a sink or bowl for 4-8 hours, depending on how firm you want your cheese to be. The longer it hangs, the drier and firmer it will become.

6. Season and Enjoy: Once the cheese has reached your desired consistency, unwrap it and season with salt. You can also mix in herbs or other flavorings at this stage.

Troubleshooting Tips

- Texture Issues: If your cheese is too runny, ensure that the milk was heated properly and that the curds were given enough time to form. Patience is key in cheese-making!

- Off-Flavors: If you notice any off-flavors, this may be due to over-fermentation or unclean equipment. Always ensure that your tools and workspace are sanitized before starting.

Flavor Variations

Chevre is incredibly versatile, and you can easily customize it to suit your taste preferences:

- Savory Chevre: Mix in minced garlic and fresh herbs like thyme or dill for a delicious savory spread that pairs well with crackers or bread.

- Sweet Chevre: For a sweeter version, incorporate honey, dried cranberries, and chopped walnuts. This makes for a delightful topping on toast or a unique addition to a cheese platter.

Homemade Feta Cheese

Feta is a versatile, crumbly cheese known for its salty tang and rich flavor. Making feta at home is a rewarding process that allows you to enjoy fresh cheese tailored to your taste. Here's a comprehensive guide to crafting your own feta cheese, including the ingredients, step-by-step process, tips for brining and preserving, and some delicious variations.

Ingredients and Process

Ingredients:

- Goat Milk: 1 gallon

- Rennet: Follow the package instructions for the appropriate amount (usually around 1/8 teaspoon for a gallon of milk).

- Starter Culture: A mesophilic culture specifically for feta cheese.

- Brine Solution: A 10% saltwater solution (1 cup of salt per 10 cups of water).

Step-by-Step Process:

1. Heat the Milk: Begin by slowly heating the goat milk in a large pot to 86°F (30°C). Stir gently to ensure even heating and prevent scorching.

2. Add Starter Culture: Once the milk reaches the desired temperature, sprinkle the starter culture over the surface of the

milk. Stir gently to incorporate it, then cover the pot and let it sit undisturbed for about 60 minutes. This allows the milk to acidify.

3. Add Rennet: After the milk has acidified, add the rennet according to the package instructions. Stir gently for about 30 seconds, then cover the pot again and let it sit undisturbed for another 30-60 minutes, or until the curds have formed and show a clean break when cut.

4. Cut the Curds: Using a long knife, cut the curds into small cubes (about 1/2 inch). Allow the curds to rest for about 10 minutes to firm up.

5. Drain the Curds: Carefully ladle the curds into a colander lined with cheesecloth. Allow the whey to drain for about 5-10 minutes. You can press the curds gently to help release more whey.

6. Press the Curds: Transfer the drained curds into a mold or container and press them down to compact the cheese.

7. Brine the Cheese: Prepare a brine solution using a 10% saltwater mixture. Submerge the cheese in the brine and let it age in the refrigerator for 5-7 days. This aging process enhances the flavor and texture of the feta.

Tips for Brining and Preserving

• Brining Solution: A 10% saltwater solution is ideal for brining feta, as it not only enhances the cheese's flavor but also extends its shelf life. Ensure the cheese is fully submerged in the brine to prevent spoilage.

- Storage: Store the feta in its brine in an airtight container in the refrigerator. This will keep it fresh and flavorful for several weeks.

Flavor Variations

Feta cheese is incredibly versatile, and you can easily customize it to suit your taste preferences:

- Marinated Feta: For added depth of flavor, marinate your feta in olive oil with herbs such as rosemary, chili flakes, or lemon zest. This not only enhances the taste but also makes for a beautiful presentation when served.

- Herbed Feta: Mix in fresh herbs like dill, oregano, or basil directly into the cheese before brining for a flavorful twist.

Homemade Ricotta Cheese

Ricotta is a quick and versatile cheese that can elevate both savory and sweet dishes. Its creamy texture and mild flavor make it a favorite ingredient in many recipes, from lasagna to desserts. Here's a comprehensive guide to making your own ricotta cheese at home, along with suggestions for how to use it in various dishes.

Quick Ricotta-Making Guide

Ingredients:

- Goat Milk: 1 gallon

- Vinegar or Lemon Juice: 1/4 cup

Step-by-Step Process:

1. Heat the Milk: Begin by pouring the goat milk into a large pot and heating it to 195°F (90°C). Stir the milk gently as it heats to prevent it from scorching on the bottom of the pot.

2. Add Acid: Once the milk reaches the desired temperature, remove it from the heat. Stir in 1/4 cup of vinegar or lemon juice. This acid will help the curds separate from the whey.

3. Curd Formation: Allow the mixture to sit undisturbed for about 10-15 minutes. During this time, you will see the curds begin to form and separate from the whey.

4. Strain the Curds: Line a colander with cheesecloth and place it over a large bowl to catch the whey. Carefully pour the curds and whey mixture into the colander. Let it drain for about 5-10 minutes, depending on how wet or dry you want your ricotta.

5. Cool and Store: Once drained, transfer the ricotta to a bowl and let it cool. You can use it immediately or store it in an airtight container in the refrigerator for up to a week.

Uses in Dishes

Ricotta cheese is incredibly versatile and can be used in a variety of dishes:

- Savory Dishes: Layer ricotta in lasagna for a creamy texture, or stuff it into ravioli for a delicious filling. It can also be added to stuffed peppers or used as a topping for pizzas.

- Sweet Dishes: Sweeten ricotta with sugar and vanilla to create a delightful filling for desserts like cannoli or cheesecake. You can also serve it with fresh fruit and honey for a simple yet elegant dessert.

- Breakfast Options: Use ricotta as a spread on toast or pancakes, or mix it into scrambled eggs for a creamy breakfast treat.

Yogurt Preparation

Yogurt is a creamy, tangy dairy product that is not only delicious but also packed with beneficial probiotics. Here's how to make it at home using goat milk.

Step-by-Step Process:

1. Heat the Milk: Begin by pouring 1 gallon of goat milk into a large pot. Heat the milk to 180°F (82°C) to kill any unwanted bacteria. Stir occasionally to prevent scorching.

2. Cool the Milk: Once the milk reaches 180°F, remove it from the heat and allow it to cool down to 110°F (43°C). This is the ideal temperature for the yogurt cultures to thrive.

3. Add Starter Culture: Mix in a yogurt starter culture or a few tablespoons of plain yogurt that contains live cultures. Stir gently to ensure the cultures are evenly distributed throughout the milk.

4. Incubate the Mixture: Pour the mixture into clean jars or a yogurt maker. Incubate at a stable temperature of 110°F for 6-12 hours. The longer you incubate, the tangier the yogurt will become. I like to use my instant pot to culture it.

5. Strain for Greek-Style Yogurt (Optional): If you prefer a thicker, Greek-style yogurt, strain the yogurt through cheesecloth or a fine mesh strainer after incubation. This will remove excess whey and create a creamier texture.

Flavor Variations:

- Fruity Yogurt: Blend in fresh fruits like berries, mango, or peaches for a refreshing flavor.

- Sweetened Yogurt: Add honey or maple syrup for sweetness, and mix in vanilla extract for extra flavor.

- Savory Yogurt: Stir in herbs and spices like dill or garlic for a savory dip or dressing.

Kefir Making

Kefir is a fermented milk drink that is slightly tangy and effervescent, known for its probiotic benefits. It's simple to make and can be enjoyed on its own or used in smoothies and salad dressings.

Step-by-Step Process:

Prepare the Milk: Start with room-temperature goat milk. You can use about 1 quart for a small batch.

1. Add Kefir Grains: Add kefir grains to the milk. Kefir grains are a combination of bacteria and yeasts that ferment the milk. Use about 1-2 tablespoons of grains per quart of milk.

2. Cover and Ferment: Cover the jar with a clean cloth or a lid (not airtight) to allow airflow. Let the mixture ferment at room temperature for about 24 hours. The fermentation time

can vary based on the temperature and your taste preference; longer fermentation will result in a tangier flavor.

3. Strain the Kefir: After fermentation, strain out the kefir grains using a fine mesh strainer. The grains can be reused for your next batch of kefir.

4. Refrigerate: Transfer the strained kefir to a clean jar and refrigerate. It can be consumed immediately or stored in the fridge for up to a week.

Flavor Variations:

- Fruit-Infused Kefir: Blend kefir with fruits like strawberries, bananas, or mango for a delicious smoothie.

- Spiced Kefir: Add a pinch of cinnamon or nutmeg for a warm, comforting flavor.

- Herbed Kefir: Mix in fresh herbs like mint or basil for a refreshing drink or salad dressing.

Goat Butter

Making goat butter at home is a rewarding process that allows you to enjoy the rich, creamy flavor of fresh butter. Here's a comprehensive guide to help you through the steps of cream separation, churning, and storing your homemade goat butter.

Cream Separation Process

To begin, you'll need to chill your goat milk. This step is crucial as it allows the cream to rise to the top, making it easier to skim off. Here's how to do it:

1. Chill the Goat Milk: Place your fresh goat milk in the refrigerator and let it sit undisturbed for about 24 hours. This resting period allows the cream to separate from the milk naturally.

2. Skim the Cream: After the milk has chilled, you'll notice a layer of cream forming on the surface. Using a ladle or a large spoon, carefully skim the cream off the top. Aim to collect as much cream as possible without disturbing the milk underneath.

Note: Goat milk doesn't separate as well as cow milk

Butter Churning Process

Once you have your cream, it's time to churn it into butter. You can use a stand mixer, a hand mixer, or even a traditional butter churn. Here's a simple method using a stand mixer:

1. Prepare the Mixer: For best results, chill the mixing bowl and whisk attachment in the refrigerator for about 30 minutes before you start. This helps the cream whip more effectively.

2. Whip the Cream: Pour the collected cream into the chilled mixing bowl. Start mixing on a low speed to avoid splattering, then gradually increase to high speed. Continue whipping until the cream thickens and eventually separates into butter and buttermilk. This process usually takes about 10-15 minutes.

3. Recognizing Butter Formation: As you churn, the mixture will first turn into whipped cream and then begin to clump together. You'll notice the buttermilk separating from the

butter. Keep churning until the butter is smooth and satiny, and the churning becomes more difficult.

Rinsing and Storing the Butter

After the butter has formed, it's important to rinse it to remove any remaining buttermilk, which can cause the butter to spoil more quickly.

1. Rinse the Butter: Place the butter in a bowl and rinse it under cold water. Use your hands to knead the butter gently, ensuring that all buttermilk is washed away. This step is crucial for extending the shelf life of your butter.

2. Knead and Store: Once rinsed, knead the butter to remove excess water. You can add salt or herbs at this stage for flavored butter. Store your homemade goat butter in an airtight container in the refrigerator. It can also be frozen for longer storage.

Flavoring Options

To enhance your goat butter, consider adding various flavors:

- Herbs: Mix in fresh or dried herbs like rosemary, thyme, or chives for a savory spread.

- Spices: Incorporate spices such as garlic powder, paprika, or black pepper for an extra kick.

- Sweet Variations: For a sweet twist, blend in honey or cinnamon, perfect for spreading on toast or pancakes.

With these steps, you can enjoy the delightful taste of homemade goat butter, perfect for cooking, baking, or simply spreading on your favorite bread!

Goat Milk Ice Cream

Making goat milk ice cream is a delightful way to enjoy the unique flavor and creamy texture of goat milk. This homemade treat is not only delicious but also a fantastic option for those who may be lactose intolerant, as goat milk typically contains lower lactose levels than cow's milk. Here's a comprehensive guide to creating your own goat milk ice cream, along with some flavor ideas to inspire your culinary creativity.

Base Recipe for Creamy Texture

Ingredients:

- 2 cups of goat milk

- 1 cup of sugar (adjust to taste)

- 4 egg yolks

- 1 teaspoon of vanilla extract

Step-by-Step Process:

1. Combine Ingredients: In a medium saucepan, whisk together the goat milk, sugar, and egg yolks. This mixture will form the base of your ice cream.

2. Cook the Mixture: Place the saucepan over medium heat and cook the mixture, stirring constantly. You want to heat it

until it thickens slightly and reaches about 170°F (77°C). Be careful not to let it boil, as this can cause the eggs to scramble.

3. Cool the Mixture: Once thickened, remove the saucepan from the heat and stir in the vanilla extract. Allow the mixture to cool to room temperature, then refrigerate it for at least 2 hours or until it is completely chilled.

4. Churn the Ice Cream: Pour the chilled mixture into an ice cream maker and churn according to the manufacturer's instructions. This process typically takes about 20-30 minutes, depending on your machine.

5. Freeze: Once churned, transfer the ice cream to an airtight container and freeze for at least 4 hours to firm up before serving.

Flavor Ideas

While the base recipe is delicious on its own, you can easily customize your goat milk ice cream with various flavors:

- Honey Lavender: Add 1/4 cup of honey and 1 tablespoon of dried lavender flowers to the base mixture before cooking. This floral twist is perfect for a refreshing summer treat.

- Chocolate Almond: Stir in 1/2 cup of cocoa powder and 1 teaspoon of almond extract to the base mixture for a rich and decadent flavor. You can also add chopped almonds for added texture.

- Strawberry Swirl: Puree fresh strawberries and swirl them into the churned ice cream just before transferring it to the freezer. This adds a fruity freshness that complements the creaminess of the goat milk.

- Cinnamon Spice: Mix in 1 teaspoon of ground cinnamon and a pinch of nutmeg for a warm, comforting flavor that's perfect for fall.

Tips for Success

- Quality Ingredients: Use high-quality goat milk and fresh eggs for the best flavor and texture.

- Chill Thoroughly: Ensure that your mixture is completely chilled before churning. This helps achieve a smoother texture in the final product.

- Storage: Store your goat milk ice cream in an airtight container in the freezer. For the best texture, allow it to sit at room temperature for a few minutes before scooping.

With this comprehensive guide, you can create delicious homemade goat milk ice cream that's sure to impress your family and friends. Enjoy experimenting with different flavors and savoring the creamy goodness of this delightful treat!

Troubleshooting and Tips for Dairy Processes

When working with goat milk and its various products, understanding how to prevent common issues and ensure proper storage is essential for achieving the best results. Here's a comprehensive guide to help you navigate the dairy-making process effectively.

Preventing Curdling or Separation

Curdling can be a frustrating issue when cooking with milk, but there are several strategies you can employ to prevent it:

1. Gentle Heating: Always heat goat milk slowly and gently. Rapid temperature changes can cause the proteins in the milk to denature and clump together, leading to curdling. Use low to medium heat and stir frequently to distribute the heat evenly.

2. Quality Ingredients: Start with fresh, high-quality goat milk. The freshness of your ingredients plays a significant role in the final product. Older milk may have a higher acidity level, which can contribute to curdling.

3. Acidity Control: Be mindful of the acidity of any additional ingredients you are using. Ingredients like lemon juice or vinegar can increase the acidity of the mixture, which may lead to curdling if not balanced properly.

4. Thickening Agents: If you're making sauces or soups, consider using a thickening agent like cornstarch or flour. Mixing these with a small amount of cold liquid before adding them to the hot mixture can help stabilize it and prevent curdling.

Storage and Shelf Life Considerations

Proper storage is crucial for maintaining the freshness and safety of goat milk products. Here are some tips to ensure your dairy items stay in top condition:

1. Airtight Containers: Always store goat milk and its products in airtight containers. This helps prevent exposure to air, which can lead to spoilage and off-flavors.

2. Refrigeration: Keep goat milk products in the refrigerator at a consistent temperature, ideally below 40°F (4°C). This helps slow down bacterial growth and extends shelf life.

3. Use Within Recommended Timeframes: Pay attention to the recommended storage times for different goat milk products. For example, fresh goat milk is best consumed within a week, while yogurt and cheese may last longer if stored properly.

4. Freezing Options: If you have excess goat milk or products, consider freezing them. Goat milk can be frozen for up to three months, while cheese and butter can also be frozen, though their texture may change slightly upon thawing.

Exploring the Versatility of Goat Milk

From tangy cheeses to creamy ice creams, goat milk adds a delightful and nutritious touch to your culinary creations. Whether you're crafting yogurt, experimenting with flavored butter, or making ice cream, each recipe offers a unique opportunity to explore the versatility of this remarkable ingredient. Embrace the process, and don't hesitate to try new flavors and techniques to enhance your dairy-making adventures!

Non-Edible Uses of Goat Milk

Goat milk is not only a culinary delight but also a remarkable ingredient in skincare and beauty products. Its rich nutrient profile, natural moisturizing properties, and soothing qualities make it an excellent base for various non-edible applications. Goat milk is commonly used in the formulation of soaps, lotions, and facial treatments, providing hydration and nourishment to the skin. Its gentle nature makes it suitable for all skin types, including sensitive skin, and can help improve skin texture and appearance. Whether incorporated into homemade beauty products or commercial formulations, goat milk serves as a versatile and beneficial ingredient in the realm of skincare.

Soap-Making with Goat Milk

Goat milk is a highly sought-after ingredient in soap-making due to its numerous skin benefits. Its naturally high-fat content provides intense moisturization, while lactic acid gently exfoliates the skin by removing dead cells, leaving it smooth and refreshed. Additionally, goat milk's pH level is close to that of human skin, which helps maintain a balanced skin barrier and reduces the risk of irritation. This makes goat milk soap an excellent choice for those with sensitive skin or conditions like eczema and psoriasis.

Cold-Process Goat Milk Soap Recipe

Making goat milk soap at home is a fun and rewarding project. Here's a comprehensive guide to creating your own cold-process goat milk soap.

Ingredients:

- 16 oz Olive Oil: Provides moisture and nourishment.

- 8 oz Coconut Oil: Adds lather and hardness to the soap.

- 8 oz Palm Oil (sustainably sourced): Contributes to the soap's firmness.

- 4 oz Shea Butter: Offers additional moisturizing properties.

- 4.5 oz Lye (sodium hydroxide): Necessary for the saponification process.

- 12 oz Goat Milk (frozen into cubes): The star ingredient for hydration and nourishment.

- Essential Oils (e.g., lavender, eucalyptus): For fragrance and therapeutic benefits.

- Natural Colorants (optional): Such as turmeric or activated charcoal for visual appeal.

Instructions:

1. Prepare Goat Milk and Lye Solution: Start by slowly adding lye to the frozen goat milk while stirring continuously. This method helps keep the milk cold and prevents scorching, which can cause discoloration. The milk may change color slightly, but this is normal.

2. Melt Oils: In a separate large pot, combine and melt the olive oil, coconut oil, palm oil, and shea butter. Once melted, allow the mixture to cool to about 100°F (38°C).

3. Combine Lye Solution with Oils: Carefully add the lye mixture to the melted oils. Use a stick blender to mix until you reach "trace," which is the point where the mixture thickens to a pudding-like consistency.

4. Add Scents and Colorants: Once you achieve trace, mix in your chosen essential oils and any natural colorants you desire. Stir well to ensure even distribution.

5. Pour and Set: Pour the soap batter into molds, smoothing the top if necessary. Cover the molds with a towel to insulate them and allow the soap to set for 24-48 hours.

6. Cure: After the soap has hardened, carefully unmold it and cut it into bars. Place the bars on a drying rack and allow them to cure for 4-6 weeks. This curing process is essential for achieving optimal hardness and quality.

Common Issues and Solutions:

- Discoloration: If the goat milk is scorched during the lye mixing process, it can turn a burnt orange or brown color. To prevent this, always use frozen goat milk and add the lye slowly.

- Curdling: To avoid curdling during mixing, ensure that both the oils and the lye solution are at similar temperatures (around 100°F) before combining them.

Skincare Products with Goat Milk

Goat milk is renowned for its nourishing qualities, making it an excellent ingredient in a variety of skincare products. Its rich composition offers hydration, gentle exfoliation, and anti-aging benefits, making it suitable for all skin types. Below, we'll explore how to create goat milk lotion, face masks, and creams, along with essential tips for packaging and preservation.

Goat Milk Lotion Recipe

Creating your own goat milk lotion at home is a straightforward process that allows you to enjoy the moisturizing benefits of goat milk. Here's a comprehensive recipe to get you started.

Ingredients:

- 8 oz Distilled Water: Acts as the base for the lotion.

- 4 oz Goat Milk: Provides hydration and nourishment.

- 2 oz Almond Oil: Adds moisture and helps to soften the skin.

- 1 oz Shea Butter: Offers additional moisturizing properties.

- 1 oz Emulsifying Wax: Helps blend the oil and water components.

- Preservative (e.g., Optiphen): Necessary for extending shelf life (follow product instructions).

- Essential Oils (optional): Such as chamomile or vanilla for fragrance.

Instructions:

1. Heat and Combine Oils: In a saucepan, melt the shea butter, almond oil, and emulsifying wax over low heat until fully melted and combined.

2. Warm Liquids: In a separate container, gently heat the distilled water and goat milk until warm but not boiling. This helps to ensure a smooth emulsion.

3. Blend: Slowly pour the warm liquid mixture into the melted oils while using an immersion blender. Blend until the mixture is fully emulsified and reaches a creamy consistency.

4. Cool and Add Preservatives: Allow the lotion to cool to room temperature. Once cooled, add the preservative and any essential oils you wish to include, mixing well.

5. Package: Transfer the lotion into sterilized pump bottles or jars for storage. Ensure that the containers are airtight to prevent contamination.

Packaging and Preservation Tips

- Use Airtight Containers: To prevent contamination and maintain the integrity of your lotion, always use airtight containers.

- Add Preservatives: Incorporating preservatives is crucial for extending the shelf life of liquid goat milk-based products, as they are more prone to spoilage

- Label Products: Clearly label your products with the date of creation and the recommended use period, which is typically 6-12 months.

Goat Milk Face Masks and Creams

In addition to lotions, goat milk can be used to create effective face masks, and creams that provide hydration and nourishment.

- Hydrating Mask: Combine 2 tablespoons of goat milk with 1 tablespoon of honey and 1 teaspoon of aloe vera gel. Mix well and apply the mask to your face. Leave it on for 15 minutes before rinsing off with warm water. This mask is excellent for moisturizing and soothing the skin.

- Anti-Aging Cream: For a nourishing anti-aging cream, mix equal parts goat milk and rosehip oil, along with a small amount of beeswax. Gently heat the mixture until the beeswax melts, then allow it to cool before storing it in a jar. This cream helps to hydrate the skin while providing essential fatty acids and antioxidants.

⤖

Chapter Nine:
Income Goats Can Provide

In recent years, goat farming has surged in popularity as a diverse and lucrative agricultural venture. Whether on a small family farm or a larger commercial operation, raising goats offers a unique opportunity to generate income through a variety of products. Goats are hardy animals that can thrive in diverse environments, from rural pastures to more confined urban settings, making them accessible to a wide range of farmers, regardless of scale. This flexibility is one of the key reasons goat farming has become a popular choice for those looking to diversify their agricultural businesses or enter into animal farming for the first time.

The appeal of goat farming lies in its versatility. Goats can be raised for a variety of purposes, including milk, meat, fiber, and even land management. By tapping into multiple income streams, goat farmers can build a resilient business model that maximizes profitability. Whether you are focused on dairy production, meat production, or niche products like goat wool, there is a market for high-quality goat products. Small-scale farmers can generate substantial income by focusing on artisanal goods, while larger operations can benefit from economies of scale and reach broader markets.

However, the success of any goat farming enterprise depends on understanding market demands and ensuring that products meet the highest standards of quality. Consumers are increasingly looking for locally-sourced, sustainable, and ethically raised products, whether they are purchasing fresh goat milk, premium meat cuts, or specialty fibers. Farmers who are attuned to these trends and prioritize quality will be able to carve out a profitable niche in a competitive marketplace.

In this chapter, we will explore the different ways in which goats can generate income, from traditional avenues like milk and meat production to more innovative opportunities such as agritourism and fiber harvesting. Each section will delve into the market demands, production practices, and best strategies for capitalizing on the diverse potential of goat farming.

Goat Milk Production

The popularity of goat milk has been steadily rising in recent years, largely due to its health benefits and suitability for individuals with lactose intolerance. Goat milk contains less lactose than cow milk, making it easier for many people to digest. It is also rich in essential nutrients such as calcium, magnesium, and vitamin A, offering a nutritious alternative to cow milk.

Additionally, goat milk is lower in fat and cholesterol, contributing to its appeal among health-conscious consumers.

One of the key advantages of goat milk over cow milk is its superior digestibility. The fat molecules in goat milk are smaller and more easily broken down by the body, which may make it more suitable for people with digestive issues or those who experience discomfort from cow milk.

Furthermore, goat milk has a naturally higher proportion of short- and medium-chain fatty acids, which are beneficial for the body's energy metabolism.

Milking goats requires daily care and attention. Goats need to be milked at least once a day, although many farmers choose to milk twice daily to increase production. The milking process should be conducted with clean, sanitized equipment to ensure the milk remains free from contaminants. Health monitoring is crucial, as goats can develop udder infections such as mastitis, which can impact milk quality. Proper nutrition and a stress-free environment are also key factors in maintaining healthy, high-yielding dairy goats.

Raw vs. Pasteurized Milk

- Raw Goat Milk: Raw goat milk has gained traction in the niche market of health-conscious and artisanal consumers who value its unprocessed nature and perceived health benefits. Proponents of raw milk claim that it contains more beneficial enzymes, probiotics, and nutrients that can be lost during pasteurization. However, raw milk is subject to strict regulations in many regions due to concerns about food safety. In some places, the sale of raw milk is heavily regulated or outright banned, so farmers must familiarize themselves with local laws before deciding to sell raw milk. For those able to sell raw milk legally, there is often a premium price attached due to its artisanal appeal and higher perceived quality.

- Pasteurized Goat Milk: Pasteurization is a process that heats the milk to a specific temperature to kill harmful bacteria, ensuring the milk is safe for consumption and extending its shelf life. While some of the beneficial enzymes and probiotics in raw milk are destroyed during pasteurization, it remains a safer option for widespread distribution, especially in retail settings. Pasteurized goat milk has a longer shelf life, making it more viable for local dairy processors who wish to distribute it to grocery stores or retail outlets. As demand for goat milk continues to grow, pasteurized milk is becoming more commonly available, and farmers may choose to partner with dairy processors for large-scale packaging and distribution.

Value-Added Dairy Products

- Goat Cheese: Goat cheese is one of the most popular and lucrative value-added products that can be made from goat milk. There is a growing market for various types of goat cheese, including soft varieties like chèvre, crumbly cheeses like feta, and aged cheeses. Goat cheese has a distinct, tangy flavor and is often considered a delicacy. Cheese production involves a careful process of curdling, draining, and aging, depending on the type of cheese being made. Farmers interested in entering the goat cheese market can differentiate themselves by offering artisanal, small-batch varieties or experimenting with flavors such as herbs, spices, or infused oils. Proper aging and packaging techniques are essential for ensuring a high-quality, marketable product.

- Goat Yogurt: Goat yogurt is becoming increasingly popular due to its probiotic content, unique taste, and digestibility. Similar to goat milk, goat yogurt has a smoother, less acidic taste compared to cow yogurt, which makes it an appealing choice for many consumers.

The yogurt-making process involves fermenting goat milk with live cultures, creating a creamy, tangy product that is rich in probiotics and beneficial for gut health. Farmers looking to produce goat yogurt can set themselves apart by offering organic, flavored, or customized yogurt varieties that cater to specific consumer preferences, such as low-sugar or lactose-free options.

- Other Dairy Products: In addition to cheese and yogurt, goat milk can be used to produce a range of other value-added products, including soaps, lotions, and other skincare items. Goat milk is prized for its moisturizing properties, making it a popular ingredient in natural skincare products. Goat milk soap, for example, is known for being gentle on the skin and can help soothe conditions like eczema or dry skin. Similarly, goat milk lotion is often marketed for its hypoallergenic and soothing qualities.

These products can be sold through farmers' markets, online platforms, or partnerships with local boutiques, providing another profitable avenue for goat farmers to explore. By diversifying into these markets, farmers can maximize the revenue potential of their goat dairy operations.

Market Demand for Goat Meat

The demand for goat meat has been steadily increasing, driven by several factors, including its reputation as a leaner, healthier alternative to beef and lamb. Goat meat is lower in fat and cholesterol, while still being rich in protein and iron, making it an appealing choice for health-conscious consumers. This shift in consumer preferences has been particularly noticeable in markets that traditionally consume goat meat, as well as among individuals seeking healthier, lower-fat options in their diets.

Globally, goat meat holds significant cultural and culinary importance, making it a sought-after commodity in various regions. In areas like the Middle East, the Caribbean, Asia, and Africa, goat meat is a staple in many diets, celebrated for its rich flavor and versatility in cooking. For farmers, this presents a unique opportunity to tap into these international markets. Ethnic communities in Western countries also contribute to the growing demand, as immigrants seek familiar products from their home countries. Farmers who can establish connections with ethnic markets and distributors are positioned to take advantage of this global appeal.

Breeding and Raising Goats for Meat

- Choosing the Right Breeds: When it comes to raising goats for meat, selecting the right breed is critical for ensuring high yields and profitability. The best meat goat breeds include the Boer, Kiko, and Spanish goats. These breeds are known for their fast growth rates, excellent feed conversion ratios, and high-quality meat yield. The Boer goat, in particular, is the most widely recognized meat breed, known for its large frame and fast weight gain. The Kiko goat is another strong contender, particularly prized for its hardiness and ability to thrive in a variety of climates. Spanish goats, known for their adaptability and low maintenance, are also a good option for small-scale meat production.

- Feeding and Healthcare: Proper nutrition is crucial for producing high-quality meat. Goats require a balanced diet that includes high-quality forage, such as pasture grass, legumes, and hay, along with supplementary grains if needed to meet their nutritional requirements. A well-balanced diet helps ensure fast growth and optimal body condition for meat production.

In addition to nutrition, maintaining herd health is essential to producing healthy, market-ready goats. Parasite management is a key component, as goats are particularly susceptible to internal parasites, which can affect their growth and overall health. Regular deworming and rotational grazing practices can help reduce parasite loads and prevent infestations. Additionally, vaccination programs to protect goats from diseases such as overeating disease (clostridial diseases) are necessary for ensuring the health of the herd and minimizing veterinary costs.

Processing and Butchering

Farmers who choose to raise goats for meat must consider how they will handle the processing and butchering stages. Some may choose to butcher goats themselves, while others may partner with licensed slaughterhouses and processors to ensure their animals are handled according to industry standards.

When processing goats, humane slaughter practices must always be a priority. It's important to follow ethical guidelines that minimize stress and ensure a quick, pain-free process for the animal. Farmers should also be aware of local, state, and federal regulations surrounding slaughter and meat processing, as they vary depending on the region and type of market (e.g., direct sales to consumers or distribution to wholesalers).

Once the goats are processed, packaging and labeling are key steps in the marketing process. Meat should be properly packaged to ensure freshness and safety during transportation and storage. Many consumers, particularly those in niche markets, value transparency in sourcing, so clear labeling that highlights the product's origin and production methods can help attract more customers.

Best Practices for Meat Production

Successful meat production involves several key practices to maximize yield and minimize costs. Some of the best practices for raising goats for meat include:

- Rotational Grazing: This practice involves moving goats between different pasture areas to give the land time to recover and maintain nutrient-rich grazing. Rotational grazing helps maintain healthy pastures, reduces overgrazing, and minimizes parasite exposure, ultimately leading to better meat quality.

- Feed Management: Balancing pasture grazing with supplemental feeding ensures that goats have access to a balanced diet, which is essential for growth and health. Providing high-quality feed supplements during periods when pasture availability is low (e.g., winter) helps goats gain weight consistently and efficiently.

- Herd Size: Maintaining an appropriate herd size is vital for efficient management. A manageable herd allows farmers to monitor individual goats more closely, ensuring that each animal receives adequate nutrition and care. It also allows farmers to optimize the available pasture and resources, reducing the risk of overgrazing or resource depletion.

To maximize meat yield and quality, goats should be finished on a high-quality pasture or supplemented feed that encourages muscle growth and fat deposition. Properly finishing goats before slaughter ensures a higher yield of tender, flavorful meat, which is essential for commanding a premium price in the market.

Angora Goats

Angora goats are prized for their production of mohair, a luxurious and versatile fiber used in high-quality textiles. Mohair is known for its softness, sheen, and durability, making it a highly desirable material in fashion and home décor. The global demand for sustainably produced mohair has been rising, as consumers increasingly prioritize eco-friendly and ethically sourced fibers. Angora goats thrive in environments with mild climates and require regular care to produce high-quality wool.

Managing Angora goats involves a few key practices, most notably shearing, which is typically done twice a year. The timing of shearing is critical for ensuring the wool is harvested at its peak quality—usually in the spring and fall. After shearing, the wool must be cleaned and processed to remove dirt and oils. The wool is then sorted and graded based on its quality, with finer fibers commanding higher prices.

For farmers interested in entering the mohair market, establishing connections with textile producers, artisans, and manufacturers is essential. Markets for raw and processed mohair are often found in niche textile industries, where demand is growing for sustainable and luxury fibers. By working with established buyers or developing a reputation for producing high-quality mohair, small farmers can tap into profitable, niche markets.

Cashmere Goats

Cashmere goats produce one of the most coveted fibers in the world: cashmere. Known for its unmatched softness and warmth, cashmere commands premium prices in the fashion industry, particularly for high-end clothing, scarves, and blankets. The global demand for cashmere continues to grow, particularly in luxury markets, with a strong emphasis on sustainably sourced fibers. Cashmere goats are typically raised in cooler climates, where the harsh weather encourages the production of the fine, insulating undercoat that makes cashmere so desirable.

Raising cashmere goats requires particular care to ensure the production of top-quality fiber. These goats must be handled gently to prevent damage to the delicate undercoat. Proper shearing techniques are essential; cashmere must be carefully separated from the coarse outer hair (guard hair) to maintain its softness and fine texture. Shearing typically occurs once a year, usually in early spring when the weather starts to warm up.

For small-scale farmers, entering the cashmere market presents both opportunities and challenges. Because of the specialized nature of cashmere production, farmers must focus on quality control and efficient processing. By offering high-quality, ethically sourced cashmere, small farmers can command a significant price premium. However, entering the market requires a solid understanding of fiber handling, care, and marketing strategies to compete with larger operations.

Other Fiber-Producing Goats

In addition to Angora and Cashmere goats, there are other breeds that produce desirable fibers, including Pygora and Nigora goats.

These breeds combine the fiber characteristics of Angora goats and Pygmy goats, offering unique textures and fiber qualities that appeal to artisans and small-scale textile producers.

- Pygora goats are known for their soft, curly fiber, which can range from fine to medium in texture. Their fiber can be used for spinning into yarn, making it popular among handcrafters and knitters.

- Nigora goats produce a fiber that is also soft and fine, blending traits of Angora and Pygmy goats. Their fiber is often used in the creation of handmade textiles, including scarves, mittens, and socks.

The appeal of these fiber-producing goats lies in their versatility and ability to provide additional or supplemental income streams for small-scale farmers. These goats can be particularly attractive to farmers looking to diversify their operations, as they often require less space and fewer resources than larger livestock operations. Additionally, the demand for unique, handspun fibers among textile artisans and craft markets offers another potential revenue stream.

While these breeds may not command the same high prices as mohair or cashmere, they present an affordable and sustainable option for farmers interested in entering the fiber market. By focusing on quality fiber production and establishing connections with niche markets, small-scale farmers can successfully compete in this growing sector.

Goat Rental for Brush Clearing

One of the lesser-known but growing revenue opportunities for goat farmers is goat rental for brush clearing. Goats are natural foragers and are particularly effective at clearing land of unwanted vegetation, such as brush, weeds, and even invasive plants like kudzu. Unlike traditional land-clearing methods, which often require costly machinery, herbicides, or manual labor, goats offer an eco-friendly and cost-effective alternative. Their ability to graze on a wide variety of plants, including shrubs and tough grasses, makes them an ideal solution for managing overgrown properties, preparing land for farming, or clearing areas prone to wildfires.

Farmers interested in starting a goat rental business can benefit from this service in several ways:

- Marketing to Landowners: Many landowners prefer using goats because it reduces the need for chemical pesticides and fertilizers, making it a greener alternative. Goats are also less disruptive than heavy machinery and can clear land without compacting the soil or disturbing wildlife habitats.

- Setting Up a Rental Operation: To begin offering goat rental services, farmers need to develop a clear pricing structure based on the size of the land to be cleared, the duration of the rental, and the number of goats required. It's also important to draft contracts that outline the responsibilities of both parties, including the duration of the rental, the level of care required, and any damage caused by the goats.

- Logistics and Herd Maintenance: A successful rental business requires managing the logistics of transporting and maintaining the herd. Farmers will need reliable transportation for the goats and ensure that they have access to sufficient water, shelter, and food while working. Regular veterinary care and maintaining herd health are critical to the success of the business.

Agritourism Ventures

Agritourism is an expanding market, especially as urban populations seek out unique, experiential activities that allow them to connect with nature and agriculture. Goat farming offers a variety of agritourism opportunities that can generate significant revenue.

- Goat Yoga: Goat yoga has become a highly popular trend, combining yoga practice with the calming presence of goats. In goat yoga sessions, participants practice yoga in a serene environment while friendly goats roam around, often interacting with the participants. This fun, lighthearted practice appeals to a broad demographic, especially urban dwellers looking for a unique experience in rural settings. The novelty of practicing yoga with goats has turned into a social media phenomenon, which can help farms attract attention and grow their customer base.

- Educational Farm Tours: Another lucrative agritourism opportunity is educational farm tours. These tours allow visitors to learn about goat farming, the sustainability of raising goats, and the benefits of different goat products like milk, cheese, and fiber. Farmers can offer tours that include live demonstrations, such as goat milking, cheese-making, and fiber harvesting, providing a hands-on learning experience for all ages. Educational farm tours are popular with schools, educational groups, families, and hobby farmers seeking to learn more about sustainable farming practices.

- Farm Stay Programs: Farm stay programs provide an immersive experience for tourists who wish to spend a night or more on a goat farm. Guests can take part in daily farm activities such as milking goats, cheese-making, or fiber production, while learning about the operations of a working farm. These programs attract individuals or families looking to experience farm life in a comfortable, engaging way. Unique lodging options, workshops, and activities can add value to the farm stay experience and generate additional income, especially during peak tourist seasons.

Goats as Pets or Breeding Stock

The market for goats as pets or breeding stock has grown substantially, especially in recent years, as more people embrace goat ownership for both practical and recreational purposes.

- Goat Sales: Goats as pets have gained popularity among hobby farmers, families, and smallholders. Breeds like Pygmy goats, Nigerian Dwarf goats, and Miniature goats are particularly sought after due to their smaller size, friendly nature, and low-maintenance care requirements. These goats are often kept as companion animals or for 4-H projects and other educational activities. Farmers can sell goats directly to buyers through local agricultural fairs, online marketplaces, or farm visits.

- Breeding Programs: Farmers can also establish breeding programs to produce goats with desirable traits, such as high milk production, superior meat yield, or high-quality fiber. Breeding goats for specific traits, such as genetic improvements for milk production or meat quality, requires careful selection and record-keeping to maintain the health and quality of the herd. Specialized breeding programs can be highly profitable, especially if the farmer develops a reputation for producing top-quality animals.

Goat farming offers a wide range of income opportunities, making it a versatile and profitable venture for farmers. From meat and milk production to fiber harvesting and agribusinesses such as goat yoga or farm stays, the potential revenue streams are diverse and flexible. These opportunities allow farmers to cater to niche markets, such as health-conscious consumers, artisans, and tourists, while leveraging the natural capabilities of goats for tasks like brush clearing.

When deciding which goat-related business model to pursue, farmers should consider their local markets, personal interests, and strengths. Whether focusing on dairy, meat, fiber, or agritourism, there are numerous ways to capitalize on the growing demand for goat products and services.

However, achieving long-term success in goat farming requires a focus on building a sustainable and well-managed operation. Maintaining herd health, implementing effective breeding and care practices, and exploring innovative marketing strategies will ensure that farmers can sustain their businesses and continue to grow in this rewarding industry. By diversifying revenue streams and adapting to market demands, goat farmers can create thriving, long-lasting enterprises.

‿⸜∘

Chapter Ten:
Conclusion

Recap of Key Insights
and Practical Aspects of Goat Keeping

Breed Selection

Selecting the appropriate breed is crucial for successful goat keeping. Each breed possesses distinct traits that cater to various objectives, whether for milk, meat, or fiber production. When choosing a breed, consider aspects such as productivity, temperament, and how well they adapt to your local environment. For instance, dairy breeds like Nubians and Saanens are renowned for their high milk yields, while Boer goats are favored for their meat quality. Fiber-producing breeds, such as Angoras and Cashmeres, provide valuable wool but necessitate more specialized care. Additionally, the temperament of the breed is important; some goats are calm and easy to manage, while others may be more spirited or independent.

Aligning your goals with the right breed ensures that your goats will thrive and meet the needs of your homestead.

Housing and Shelter

Providing proper housing is vital for the safety, health, and comfort of your goats. Although goats are resilient animals, they still require protection from harsh weather, predators, and parasites. When designing their shelter, ensure there is ample space for each goat to move freely. Good ventilation is essential to prevent respiratory problems, particularly in humid or extremely cold climates. Protecting against predators like dogs, coyotes, and foxes is crucial, so the shelter should be secure with robust fencing and gates. The living area must be clean, dry, and draft-free, allowing goats to rest comfortably. A well-designed shelter not only safeguards your goats but also enhances their overall well-being and productivity.

Feeding and Nutrition

Nutrition plays a pivotal role in maintaining the health and productivity of your goat herd. A balanced diet is essential for strong immune systems, healthy growth, and reproductive success. As ruminants, goats primarily rely on forage, such as grass or hay, to fulfill their nutritional requirements. However, they also benefit from supplemental feeds, including grains, minerals, and vitamins. Always provide access to fresh, clean water, as dehydration can lead to serious health issues. A proper diet should include quality grazing opportunities, which support digestive health and prevent boredom. Regularly monitoring your goats' weight, body condition, and overall health will help ensure they receive the necessary nutrients for optimal growth and productivity.

Health Management

Routine health management is essential for keeping your goats in excellent condition. Regular health checks can help identify potential issues early, whether they involve parasites, infections, or injuries. Vaccinations are critical for preventing common diseases such as tetanus, pneumonia, and caseous lymphadenitis (CL). Managing parasite control is also vital, as goats are particularly vulnerable to internal parasites like worms. Collaborate with a veterinarian to create a health program tailored to your herd's specific needs. Establishing a good relationship with a veterinarian experienced in goat care is invaluable, as they can provide guidance on routine care, address health concerns, and offer advice for emergencies. Preventative care is always more effective than addressing problems after they arise.

Breeding

Breeding is an exciting and rewarding aspect of goat keeping, but it requires careful planning and knowledge. Understanding the reproductive cycle of goats is key to successful breeding. It typically comes into heat every 18 to 21 days, and breeding should occur during these cycles. When selecting breeding stock, prioritize traits such as health, conformation, and productivity. Choose bucks and does that complement each other's strengths and weaknesses to enhance the overall quality of the herd. During pregnancy, they need extra care, particularly in the final trimester, as they prepare to give birth. Monitor their health closely, provide a nutritious diet, and ensure they have a quiet, clean space for delivery. After birth, focus on the care of the newborn kids, ensuring they receive adequate colostrum and monitoring them for any signs of illness. A successful breeding program will help ensure a sustainable, growing herd, providing both milk and meat for your homestead.

The Rewards of Raising Goats: A Fulfilling and Productive Venture

Raising goats is not just a practical endeavor for generating income or providing food; it is an enriching experience that profoundly impacts one's sense of purpose, personal satisfaction, and connection to nature. While goats offer valuable products such as milk, cheese, meat, and fiber, the true rewards often extend beyond mere financial gain.

Personal Fulfillment Through Self-Sufficiency

There is a unique pride that comes from being able to provide for yourself and your loved ones through your own efforts. The milk from your goats can be transformed into fresh cheese, yogurt, or even soap—products that nourish both body and spirit while adding value to your homestead. Raising goats fosters self-reliance, reducing dependence on store-bought goods and deepening your connection to the food you consume. This self-sufficiency is not just practical; it empowers you to reclaim control over your life and well-being. For many, the ability to produce something from the land is a source of unparalleled pride, infusing life with renewed purpose.

The joy of crafting your own cheese or butter from goat milk, creating warm clothing from their fiber, or knowing that the meat from your goats nourishes your family is a fulfillment that is hard to replicate in a conventional lifestyle. Contributing directly to your household's needs creates a profound sense of satisfaction, as your efforts are woven into every meal and cherished moments spent with family.

The Emotional Satisfaction of Caring for Goats

Beyond tangible rewards, raising goats provides deep emotional satisfaction. Caring for animals demands commitment, patience, and empathy. Goats are curious, intelligent, and social creatures, and building a bond with them can be a deeply rewarding experience. Their unique personalities shine through, whether it's a playful kid exploring its surroundings or a doe affectionately seeking attention. This interaction fosters companionship and connection that enriches your life beyond the routine of animal care.

Nurturing your goats and witnessing their growth and productivity is immensely gratifying. The cycles of life on your farm—from the birth of healthy kids to the satisfaction of milking a doe after months of care—bring joy and fulfillment. Each milestone, from successful breeding to watching your goats graze contentedly, is a triumph that enhances your sense of accomplishment and connection to the land.

Building Resilience Through Challenges and Triumphs

Raising goats comes with its challenges, but overcoming these hurdles fosters personal growth and resilience. From managing health issues like parasites to navigating difficult births, goat keeping teaches perseverance and problem-solving. These challenges test your adaptability and growth, both as a goat keeper and as an individual. Each obstacle overcome builds confidence, and every success, no matter how small, reinforces the resilience gained from tackling tough situations.

Moreover, goat-keeping offers opportunities for introspection and learning. The quiet moments spent in the barn or field with your herd provide peace and clarity. Daily routines, from feeding to milking, create moments of mindfulness, allowing you to reflect on your journey and growth as a homesteader. The natural cycles of your animals—pregnancy, birth, and growth—mirror the ebb and flow of life itself, reminding you of both the fragility and strength inherent in nature.

Through goat keeping, you also learn to manage the unpredictable nature of farming. Weather conditions, health setbacks, and market fluctuations present challenges that require flexibility and determination. Yet, it is these very challenges that make success so rewarding. Each step forward, whether mastering a new technique or achieving a successful breeding, adds a layer of resilience applicable to all areas of life.

The Joy of Sharing with Others

Another profound aspect of raising goats is the ability to share the fruits of your labor with others. Whether gifting homemade cheese to a neighbor, selling goat milk products at a local market, or sharing stories about your experiences, goat keeping fosters community and connection. This sense of sharing extends to both tangible goods and intangible experiences—helping others through your knowledge and offering a glimpse into your lifestyle.

The relationships formed through goat keeping, whether with fellow homesteaders, consumers, or those who benefit from your goods, provide a sense of belonging and purpose. The simple act of giving something valuable, whether fresh goat milk or handspun yarn, contributes to the well-being of those around you and strengthens the fabric of your community.

A Deepened Connection to Nature and the Land

Ultimately, raising goats offers more than just the satisfaction of completing daily chores; it fosters a deepened connection to the land and the natural world. The rhythms of the seasons, the care for the animals, and the upkeep of the land create a harmonious relationship that enriches your understanding of sustainability. Goats are excellent foragers, and their ability to clear brush and manage land exemplifies how humans can work in partnership with nature to create sustainable ecosystems.

In this way, goat keeping transcends being merely a hobby or a means of productivity; it becomes a lifestyle that aligns with nature's cycles, contributing to a more balanced and sustainable way of living. This connection to the land enhances well-being, providing a sense of rootedness and purpose that transcends the fast-paced modern world. The satisfaction of working with the land, witnessing the fruits of your labor in healthy, happy goats, and living in harmony with nature creates a profound sense of fulfillment.

Lifelong Learning and Adaptation

Embracing the journey of lifelong learning is one of the most rewarding aspects of goat keeping. Each year with goats brings new experiences, challenges, and insights that deepen your understanding of these remarkable animals and the complexities of managing a goat farm. Whether you are a novice or an experienced keeper, there is always more to learn, and the landscape of goat care is continually evolving. By embracing this process of growth and adaptation, you can ensure that your herd remains healthy and your farm thrives.

The Ever-Changing Nature of Goat Keeping

Raising goats is a dynamic endeavor. Each season, kidding cycle, and change in weather or environment presents new challenges. What works well one year may not be as effective the next, and conditions may shift in ways that require fresh strategies and approaches. This variability is part of the adventure and beauty of farming. The more you learn and adapt, the better equipped you become to anticipate and address these changes.

Over time, your relationship with your herd will become more intuitive as you attune yourself to their behaviors, needs, and subtle signs of illness or distress. However, it's important to acknowledge that even seasoned goat keepers encounter setbacks and surprises. Mistakes are an inevitable part of the learning process; they are not failures but rather opportunities for growth. Whether it's misjudging the timing of breeding, struggling with a new feeding regimen, or facing an unforeseen health issue, each mistake can impart valuable lessons that enhance your overall practice.

The Power of Community and Shared Knowledge

One of the most effective ways to navigate the challenges of goat keeping is by connecting with others who share your passion. The goat-keeping community is vast and diverse, filled with individuals who possess a wealth of knowledge and experience. Joining local goat-keeping groups, attending workshops, or participating in online forums allows you to exchange ideas, troubleshoot issues, and share best practices with like-minded individuals.

Being part of this community provides a support network where you can ask questions, seek advice, and celebrate successes together. It fosters camaraderie as you realize that every goat keeper has faced similar challenges and joys. This sense of solidarity and mutual support is invaluable, reinforcing the idea that you are not alone on your journey. Whether you're seeking tips on breeding, health management advice, or simply a space to share your triumphs, these connections help you grow as a goat keeper and encourage continuous learning.

Staying Informed and Adapting to New Advancements

As the field of goat care continues to evolve, staying informed about the latest advancements in breeding techniques, health management, and sustainable farming practices is essential. Ongoing research aims to improve the quality of care for goats, and new innovations can provide valuable tools for enhancing the health of your herd and the sustainability of your farm. For example, there are now more efficient and humane parasite control methods, advances in genetic breeding for better milk production, and improved feed options that promote healthier, more productive goats.

Attending conferences, subscribing to relevant journals, or following respected experts in the field can help you stay updated on these developments. By continually seeking new information and incorporating it into your practices, you can optimize the care you provide for your goats, improve the overall health of your herd, and enhance the sustainability of your farm. This ongoing commitment to learning and adaptation is a hallmark of a successful goat keeper and ensures the long-term success of your homestead.

Improving Herd Health and Farm Sustainability

As you expand your knowledge, you will begin to recognize the subtle ways in which your decisions impact your herd's health and the sustainability of your farm. For instance, improvements in your feeding practices can lead to stronger, more resilient goats that are less susceptible to illness and produce higher-quality products. Similarly, refining your breeding practices over time can result in healthier, more productive offspring with better temperaments and disease resistance.

Additionally, adapting your farming techniques to align with sustainable practices can contribute to a more environmentally friendly operation. Implementing rotational grazing to prevent overgrazing, using organic fertilizers, or building energy-efficient housing are all ways to adopt sustainable farming principles. These practices help ensure that your farm is not only productive but also responsible and resilient in the face of changing environmental conditions.

The Growth Mindset in Goat Keeping

The key to long-term success as a goat keeper lies in adopting a growth mindset. This means viewing challenges not as obstacles but as opportunities to improve, learn, and refine your practices. It involves being open to trying new things, whether it's a different breed, a new milking method, or an alternative approach to breeding. It means learning from both your successes and failures and continuously striving to do better for your goats and your farm.

Developing a Connection to the Land and Animals

Raising goats transcends mere practicality; it is a deeply emotional and often spiritual journey that fosters a profound connection between the keeper, the animals, and the land they inhabit. As you dedicate time to caring for your goats—feeding them, tending to their health, and observing their behaviors—you begin to understand them on a level that goes beyond routine management. This bond strengthens with each passing day, transforming simple farm tasks into moments of connection and reflection.

The Emotional Bond with Goats

Goats, with their unique personalities and intelligence, have a remarkable ability to endear themselves to their keepers. Each goat has its own character—some may be playful and curious, while others are more reserved or independent. As you learn their individual quirks and tendencies, a relationship of trust and affection develops. Over time, they come to recognize you not just as their caregiver but as a familiar presence, greeting you with enthusiasm and seeking your attention.

This emotional bond is one of the most rewarding aspects of goat keeping. Routine chores become opportunities for personal engagement with the animals. The act of bottle-feeding a newborn kid, assisting a doe through a challenging kidding, or simply sitting quietly in the pasture while the herd grazes start to become a source of joy and fulfillment. These moments remind you of life's simple pleasures and the value of nurturing living beings.

Fostering Patience, Responsibility, and Empathy

Goat-keeping imparts invaluable life lessons, particularly in patience, responsibility, and empathy. Goats require consistent care and attention, and their well-being hinges on your dedication and reliability. Learning to meet their needs—sometimes under challenging conditions—cultivates a sense of responsibility that extends beyond the farm.

Patience is another essential quality that goat keeping nurtures. Animals do not adhere to human schedules, and things rarely unfold as planned. You may find yourself waiting for hours during a difficult labor or spending days nursing a sick goat back to health. These experiences teach you to slow down, be present, and accept life's unpredictability with grace.

Empathy flourishes through the experience of caring for animals. Observing your goats' behaviors, recognizing their needs, and responding to their distress fosters a deeper sense of compassion. This empathy often extends beyond the farm, influencing how you relate to others in your community and the natural world.

A Harmonious Way of Life

One of the most transformative aspects of goat keeping is how it intertwines your life with the rhythms of nature. Caring for goats requires you to be attuned to the seasons, weather, and natural cycles of life and growth. You become more connected to the land, observing its changes throughout the year and learning to work with it rather than against it.

This harmonious relationship with nature fosters a sense of balance and fulfillment. The work of tending to the goats—feeding them, maintaining their shelter, rotating pastures—becomes part of the natural ebb and flow of farm life. Each task, no matter how small, contributes to the health and sustainability of the land and the animals, creating a sense of purpose and accomplishment.

As you become more in tune with this rhythm, your perspective on work and life may shift. The line between work and leisure blurs, and the satisfaction of a job well done becomes its own reward. You begin to appreciate the interconnectedness of all living things and the importance of living in harmony with the environment.

Nurturing a Sustainable Future

Goat keeping also paves the way for a more sustainable and self-sufficient lifestyle. By raising your own animals for milk, meat, or fiber, you reduce reliance on external resources and become more resilient in the face of economic or environmental challenges. The land you steward becomes a source of nourishment and security, while the goats you care for become partners in a shared journey toward sustainability.

This sustainable way of life fosters a deep sense of gratitude for the land and the animals that sustain you. Each day presents new opportunities to give back to the land, whether through responsible grazing practices, soil conservation, or planting forage crops that benefit both the goats and the ecosystem. In return, the land rewards you with its abundance, creating a cycle of care and renewal that benefits both the keeper and the environment.

A Legacy of Connection and Stewardship

Ultimately, the connection you develop with the land and animals through goat keeping becomes a legacy of stewardship. It is a way of life that you can pass down to future generations, teaching them the value of hard work, compassion, and living in harmony with nature. This legacy encompasses not just the skills of goat keeping but also the mindset and values that accompany it—a deep respect for the land, a commitment to sustainability, and an understanding of the interconnectedness of all life.

In this way, goat keeping evolves into more than a practical endeavor or a source of income. It becomes a way of living that enriches your life and the lives of those around you. It teaches you to find joy in simple, everyday moments and to appreciate the beauty and wonder of the natural world. Through this journey, it offers a path to a more meaningful, fulfilling, and harmonious existence.

A Motivational Message for Aspiring Goat Keepers

If you've made it this far in the book, you've likely felt inspired by the possibilities that goat keeping can bring to your life. You might be bubbling with excitement at the thought of starting your own herd, but perhaps you're also experiencing a bit of uncertainty. Stepping into something new—especially something as involved as raising livestock—can indeed feel intimidating.

But here's the truth: you have what it takes to succeed as a goat keeper. With preparation, dedication, and a willingness to learn, you can transform this dream into a rewarding reality. Remember, no one starts out as an expert. Every successful goat keeper you've heard about or met began their journey just like you—with questions, doubts, and a desire to try something new.

Embrace the First Step

The hardest part of any new venture is often taking that first step. Whether it's purchasing your first goats, building a shelter, or simply committing to the idea, that initial action sets everything in motion. Don't wait until you feel 100% ready or have every detail figured out—because the truth is, you never will. Goatkeeping, like any worthwhile endeavor, is a continuous journey of learning and growth.

Start small. Begin with a manageable number of goats and gradually expand your herd as you gain confidence and experience. Each day will present new challenges, but each challenge is an opportunity to learn and grow.

You Don't Need All the Answers

It's easy to feel overwhelmed by the vast amount of knowledge required—feeding schedules, health management, breeding techniques, and more. But don't let this overwhelm hold you back. You don't need to have all the answers right now. What matters is your willingness to seek out those answers, whether through research, trial, and error, or by reaching out to experienced goat keepers.

Remember, no one becomes a seasoned goat keeper overnight. Mistakes will happen, and that's perfectly okay. Each misstep is a valuable lesson that brings you closer to success. With each passing day, you'll find yourself becoming more confident, capable, and knowledgeable.

Real Stories of Success

Many experienced goat keepers started with little to no experience, yet they thrived through determination and a willingness to learn. Take Sarah, for example, a suburban mother who knew nothing about livestock but wanted to provide fresh milk for her family. She began with two dairy goats and a borrowed book on goat care. The first year was filled with mishaps—feeding mistakes, a fence that didn't hold, and a difficult kidding. But she persevered, learning from each challenge and seeking advice from local farmers. Today, Sarah runs a thriving small dairy farm, producing milk, cheese, and soap.

Or consider James, a retired teacher who moved to the countryside with a dream of self-sufficiency. With no prior experience, he started with a small herd of goats to manage his overgrown pasture. What began as a hobby turned into a passion, and James now mentors new goat keepers in his community, sharing the lessons he learned through years of trial and error.

These stories—and countless others—demonstrate that success is possible, even if you start with nothing but a dream and a willingness to work hard.

Building Your Community

One of the most powerful resources you'll have on this journey is the goat-keeping community. From online forums and social media groups to local agricultural clubs and extension services, there are countless ways to connect with others who share your passion. Don't hesitate to ask questions, seek advice, and learn from the experiences of others. The knowledge and support you gain from this network can make a world of difference as you navigate the challenges of goat keeping.

Your Journey Awaits

As you stand on the brink of this new adventure, remember that the rewards of goat-keeping extend far beyond milk, meat, or fiber. It's a journey of personal growth, resilience, and connection to the land and animals. Each day will bring new experiences, along with new joys and lessons.

So, take that first step with confidence. Equip yourself with knowledge, embrace the challenges, and trust in your ability to succeed. The path of a goat keeper is not always easy, but it is deeply rewarding. In time, you'll look back and realize that every moment of effort was worth it—for the goats you care for, the land you nurture, and the life you build along the way.

Goats as a Path to Enriching Homestead Life

Goats offer immense value to a homestead that goes far beyond their tangible contributions of milk, meat, fiber, and offspring. They provide a unique combination of productivity, ecological benefits, and emotional fulfillment, making them an integral part of a thriving, self-sufficient lifestyle. When thoughtfully integrated into the fabric of homestead life, goats can transform not only the landscape but also the homesteader's relationship with nature, community, and sustainability.

Ecological Sustainability and Soil Health

One of the most significant yet often overlooked contributions of goats is their positive impact on soil health. Their manure, rich in nitrogen, phosphorus, and other essential nutrients, acts as a natural fertilizer, rejuvenating the soil and enhancing its fertility. Unlike chemical fertilizers, goat manure improves soil structure over time, promoting better water retention and supporting the growth of healthy plants and crops.

Additionally, goats are natural grazers and browsers. They help control invasive plant species, manage overgrowth, and contribute to pasture rejuvenation by evenly grazing and distributing seeds through their manure. This grazing behavior can enhance biodiversity, encouraging the growth of native plants and fostering a balanced ecosystem on the homestead.

Companionship and Joy

Goats are more than just livestock; they are companions. Each goat possesses a distinct personality, ranging from curious and adventurous to calm and affectionate. Their playful antics and loving nature bring joy and vibrancy to the homestead, transforming daily tasks like feeding, milking, and caring for them into sources of happiness rather than chores.

The bond formed between a goat keeper and their herd can be profoundly rewarding. Over time, you come to know each goat individually—their quirks, preferences, and behaviors. This connection fosters empathy, patience, and a sense of responsibility, enriching the emotional well-being of the keeper and creating a relationship built on mutual trust and care.

Building Community and Local Networks

Goat keeping also has the potential to strengthen community ties. Homesteaders often find themselves sharing knowledge, experiences, and resources with others who keep goats or are interested in starting their own herds. This exchange of information—whether through local farming groups, social media communities, or neighborhood networks—builds a sense of camaraderie and mutual support.

Trading goat products such as milk, cheese, soap, or fiber can further enhance community connections. Participating in local farmers' markets, bartering goods with neighbors, or collaborating on homestead projects creates a network of like-minded individuals who share a common goal of self-sufficiency and sustainable living. These interactions not only provide economic benefits but also foster a sense of belonging and shared purpose.

A Gateway to Living in Harmony with Nature

At its core, goat keeping serves as a gateway to living in greater harmony with nature. By raising goats, homesteaders engage in a sustainable cycle of life that honors the land, the animals, and the environment. Goats remind us of the interconnectedness of all living things—their health and well-being are directly tied to the health of the land they graze, the water they drink, and the care they receive.

This harmonious relationship encourages homesteaders to adopt sustainable practices, such as rotational grazing, composting, and water conservation. By doing so, they create a resilient and self-sufficient homestead that not only meets their immediate needs but also preserves and enhances the land for future generations.

A Lifestyle of Purpose and Resilience

Ultimately, raising goats is about more than productivity; it's about cultivating a lifestyle of purpose, resilience, and connection. It's about waking up each day with a sense of responsibility and fulfillment, knowing that your efforts contribute to something greater than yourself. It's about learning to live in rhythm with the natural world, finding joy in simple tasks, and embracing the challenges and rewards that come with homestead life.

Goats, with their unique blend of utility and personality, have the power to transform a homestead into a thriving, sustainable ecosystem. They enrich the soil, nourish the body, and uplift the spirit, offering a way of life that is deeply rooted in harmony, community, and respect for the earth.

Parting Words: Empowering Readers to Succeed

As you reach the conclusion of this book, I want to take a moment to express my heartfelt gratitude. Thank you for your curiosity, dedication, and willingness to explore the world of goat keeping. Whether you are a seasoned homesteader or just beginning to dream of raising goats, your commitment to learning and growing is truly commendable. This journey requires patience, resilience, and an open heart—but it also promises countless rewards.

The path ahead may not always be smooth. You will encounter challenges and uncertainties, moments when you question your abilities or face setbacks that test your resolve. But remember this: you are capable. Within you lies the strength, resourcefulness, and determination to succeed. Every challenge presents an opportunity to learn, adapt, and grow. With each step forward, you are building a life that is not only sustainable but also deeply fulfilling.

Embrace this journey with enthusiasm. Approach each day with the understanding that goat keeping is more than just a task—it is a lifestyle that connects you to the land, your animals, and your community. Celebrate the small victories, whether it's a successful kidding season or the joy of producing your own cheese. Find happiness in the daily rhythms of farm life and take pride in the progress you make, no matter how small.

Remember, you are not alone. Across the globe, countless others are walking this path, facing similar challenges and sharing in the same triumphs. Lean on this community for support, guidance, and inspiration. Seek out fellow goat keepers, join local or online networks, and never hesitate to ask questions or share your experiences. Together, we are stronger, and together, we can create a future where sustainable, responsible goat-keeping thrives.

As you move forward, I leave you with a message of hope and encouragement. The work you do with your goats has the potential to impact more than just your homestead—it can enrich your community, contribute to the health of the environment, and foster a way of life that is in harmony with nature. Your dedication to this craft reflects your vision and values, leaving a lasting legacy for future generations.

So, take that first step—or the next step—with confidence. Believe in your ability to create a thriving, sustainable, and rewarding life with goats. You possess the knowledge, passion, and perseverance to succeed. Most importantly, remember that this journey is yours to shape, filled with endless possibilities and the promise of a life well-lived.

Closing Thought:
A Legacy of Sustainability and Connection

As we conclude this journey together, I invite you to reflect on the legacy you are creating—not just for yourself but for the land, your community, and future generations. Goat keeping is more than a practical endeavor; it is a commitment to a way of life that values sustainability, connection, and stewardship. Every decision you make in raising your goats contributes to a broader vision of harmony between humans, animals, and the environment.

Imagine your homestead in the years to come. Envision a place where goats are not merely livestock but integral partners in nurturing the land. Their grazing regenerates the soil, their manure enriches your garden, and their presence infuses life and energy into your farm. In this ecosystem, nothing is wasted, and every element supports and enhances the others.

Beyond the physical benefits, raising goats fosters a profound sense of connection. It teaches patience, resilience, and empathy, reminding us of our interdependence with the natural world. It strengthens community ties through shared knowledge, collaborative efforts, and the exchange of goat-related products. In many ways, your homestead becomes a beacon of sustainability and a testament to the rewards of living in balance with nature.

This legacy extends far beyond your immediate surroundings. As you care for your goats and cultivate your land, you contribute to a global movement toward more sustainable and regenerative practices. Your efforts inspire others to consider how they, too, can live more sustainably—whether by raising their own livestock, supporting local farmers, or making choices that honor the earth.

Ultimately, goat keeping is about more than productivity; it's about creating a life rooted in purpose, connection, and sustainability. It's about finding joy in the everyday rhythms of farm life and recognizing that each act of care, no matter how small, contributes to something much larger.

As you look to the future, remember that the legacy you build is not just about the present—it's about the impact you leave on those who follow. Your homestead, your herd, and your way of life have the potential to inspire and sustain future generations, demonstrating that it is possible to live in harmony with the earth while finding fulfillment and purpose in your work.

Thank you for choosing this path and for embracing the journey of goat keeping. May it bring you countless rewards, both tangible and intangible and may your legacy of sustainability and connection continue to grow and thrive for years to come.